T0191649

Science, Technology and Medicine in Modern History

General Editor: **John V. Pickstone**, Centre for the History of Science, Technology and Medicine, University of Manchester, England (www.man.ac.uk/CHSTM)

One purpose of historical writing is to illuminate the present. At the start of the third millennium, science, technology and medicine are enormously important, yet their development is little studied.

The reasons for this failure are as obvious as they are regrettable. Education in many countries, not least in Britain, draws deep divisions between the sciences and the humanities. Men and women who have been trained in science have too often been trained away from history, or from any sustained reflection on how societies work. Those educated in historical or social studies have usually learned so little of science that they remain thereafter suspicious, overawed, or both.

Such a diagnosis is by no means novel, nor is it particularly original to suggest that good historical studies of science may be peculiarly important for understanding our present. Indeed this series could be seen as extending research undertaken over the last half-century. But much of that work has treated science, technology and medicine separately; this series aims to draw them together, partly because the three activities have become ever more intertwined. This breadth of focus and the stress on the relationships of knowledge and practice are particularly appropriate in a series which will concentrate on modern history and on industrial societies. Furthermore, while much of the existing historical scholarship is on American topics, this series aims to be international, encouraging studies on European material. The intention is to present science, technology and medicine as aspects of modern culture, analysing their economic, social and political aspects, but not neglecting the expert content which tends to distance them from other aspects of history. The books will investigate the use and consequences of technical knowledge, and how it was shaped within particular economic, social and political structures.

Such analyses should contribute to discussions of present dilemmas and to assessments of policy. 'Science' no longer appears to us as a triumphant agent of Enlightenment, breaking the shackles of tradition, enabling command over nature. But neither is it to be seen as merely oppressive and dangerous. Judgement requires information and careful analysis, just as intelligent policy-making requires a community of discourse between men and women trained in technical specialities and those who are not.

This series is intended to supply analysis and to stimulate debate. Opinions will vary between authors; we claim only that the books are based on searching historical study of topics which are important, not least because they cut across conventional academic boundaries. They should appeal not just to historians, nor just to scientists, engineers and doctors, but to all who share the view that science, technology and medicine are far too important to be left out of history.

Titles include:

Roberta E. Bivins
ACUPUNCTURE, EXPERTISE AND CROSS-CULTURAL MEDICINE

Roger Cooter
SURGERY AND SOCIETY IN PEACE AND WAR
Orthopaedics and the Organization of Modern Medicine, 1880–1948

David Edgerton
ENGLAND AND THE AEROPLANE
An Essay on a Militant and Technological Nation

Jean-Paul Gaudilliére and Ilana Löwy (*editors*)
THE INVISIBLE INDUSTRIALIST
Manufacture and the Construction of Scientific Knowledge

Thomas Schlich
SURGERY, SCIENCE AND INDUSTRY
A Revolution in Fracture Care, 1950s–1990s

Eve Seguin (*editor*)
INFECTIOUS PROCESSES
Knowledge, Discourse and the Politics of Prions

Crosbie Smith and Jon Agar (*editors*)
MAKING SPACE FOR SCIENCE
Territorial Themes in the Shaping of Knowledge

Stephanie J. Snow
OPERATIONS WITHOUT PAIN
The Practice and Science of Anaesthesia in Victorian Britain

Carsten Timmermann and Julie Anderson (*editors*)
DEVICES AND DESIGNS
Medical Technologies in Historical Perspective

Science, Technology and Medicine in Modern History
Series Standing Order ISBN 978-0-333-71492-8 hardcover
Series Standing Order ISBN 978-0-333-80340-0 paperback
(*outside North America only*)

You can receive future titles in this series as they are published by placing a standing order.
Please contact your bookseller or, in case of difficulty, write to us at the address below with
your name and address, the title of the series and one of the ISBNs quoted above.

Customer Services Department, Macmillan Distribution Ltd, Houndmills, Basingstoke,
Hampshire RG21 6XS, England

Devices and Designs

Medical Technologies in Historical Perspective

Edited by Carsten Timmermann and
Julie Anderson
Centre for the History of Science, Technology and Medicine
The University of Manchester

First published 2006 by
PALGRAVE MACMILLAN
Houndmills, Basingstoke, Hampshire RG21 6XS and
175 Fifth Avenue, New York, N.Y. 10010
Companies and representatives throughout the world

PALGRAVE MACMILLAN is the global academic imprint of the Palgrave Macmillan division of St. Martin's Press, LLC and of Palgrave Macmillan Ltd. Macmillan® is a registered trademark in the United States, United Kingdom and other countries. Palgrave is a registered trademark in the European Union and other countries.

ISBN 978-1-349-54091-4   ISBN 978-0-230-28640-5 (eBook)
DOI 10.1007/978-0-230-28640-5

This book is printed on paper suitable for recycling and made from fully managed and sustained forest sources.

A catalogue record for this book is available from the British Library.

Library of Congress Cataloging-in-Publication Data

Devices and designs : medical technologies in historical perspective / edited by Carsten Timmermann and Julie Anderson.
     p. cm.
Includes bibliographical references (p.   )
ISBN 978-0-333-80340-0 (cloth)
    1. Medical instruments and apparatus–History. 2. Medical technology–History. I. Timmermann, Carsten, 1966– II. Anderson, Julie, 1965–

R856.A5D43 2006
610.28–dc22   2006047611

10   9   8   7   6   5   4   3   2   1
15   14   13   12   11   10   09   08   07   06

Transferred to Digital Printing in 2008

Contents

List of Figures

Foreword

Thomas P. Hughes

The University of Manchester conference about technology and medicine where the papers in this volume were first presented focused upon a relationship that will surely attract the attention of many more historians of technology and historians of medicine. The interaction of technology and medicine will undoubtedly encourage many historians to designate themselves as historians of technology *and* medicine. The field is ripening rapidly and cries out for professional and public understanding. Physicians, engineers, and scientists, as well as the public, should better understand the relationship of technology and medicine, so that they can garner insights that will help them take action, improve practice, and make policy.

More than 30 years ago, the proliferation of costly medical machines led Arnold Relman, the influential editor of the *New England Journal of Medicine*, to refer to a 'medical-industrial complex,' thus playing upon the public's and the medical profession's familiarity and critical stance toward the military-industrial complex.[1] Because the complex deployed advanced and enormously expensive technology, Relman argued that costly technology would shape the future of health care. Others familiar with the health system subsequently alluded in a similar vein to a 'medical arms race' discernible in the 1970s. Some economists contended technological change was the major factor contributing to rising health care costs, adding billions to the costs.[2] Cost-driving technologies emerging in the 60s and 70s included kidney dialysis, open-heart surgery, organ transplants, and computerized tomography (CT) scans. By 1980, the United States on a per capita basis had more CT scans, open-heart surgery suites, and other costly technologies than any other nation. Since then medical machines have proliferated.

Today so much emphasis and attention is given to pharmaceutical companies and their products that the media and the public tend to overlook manufacturers of costly medical machines. A division of General Electric, the Milwaukee-based General Electric Medical Systems Division (GEMS), is the world's leading manufacturer of diagnostic imaging equipment, including digital X-ray (CT), magnetic resonance (MR), nuclear medicine (NM) imaging, and ultrasound machines.

GEM's major competitors are Dutch-based Philips, and Germany-based Siemens. By 2002, GEMS parent General Electric had the world's largest market capitalization (US$400 billion).[3]

Today the increased importance of the study of hereditary traits and potential abnormalities of cells and cell function and related therapies (genomics) encourages the collaboration of imaging equipment manufacturers and pharmaceutical companies. Such alliances will enhance the role of the machine manufacturers in healthcare. In addition, the manufacturers will have to add expertise in biomedical sciences to their established expertise in engineering and physics. In view of the role of machine builders in rising healthcare costs, historians of technology and medicine should delve more deeply into their past and present. I have made a preliminary study of medical machine manufacturers by drawing analogies between the history of electrical machine companies and medical machine companies, especially General Electric.[4]

Several essays in the Devices and Designs volume deal with machines, the growing importance of hospital staffs that maintain the machines, and medical inventions and innovations. For instance, Takahiro Ueyama and Christophe Lécuyer in their essay deal with the omnipresence of machine technology in hospitals, such as CT and MR scanners. As a result many patients feel alienated from doctors' personal care. In response, designers and manufacturers have attempted to personalize the machines by providing a friendlier interface.

The role of inventors and innovators promises to be another major growth area in the history of technology and medicine. In the United States the National Inventors Hall of Fame, which has long inducted major inventors past and present, has, in recent years, bestowed the accolade on an increasing number of inventors in the medical field. Raymond Damadian has been cited for an MR scanner, Wilson Greatbatch for a pacemaker, Dean Kamen for a medication injection device, Robert Ledley for a diagnostic X-ray system, Michael Mirowski, Morton Mower, M. Stephen Heilman, and Alois Langer for method and apparatus for monitoring heart activity, and Rangaswamy Srinivasan James Wynne, and Samuel Blum for ultraviolet surgical and dental procedures. The US National Academy of Engineering has recently introduced the Russ Prize for bioengineering achievement. Winners have been Willem J. Kolff for his pioneering work on artificial organs and Earl E. Bakken and Greatbatch for their independent development of the implantable cardiac pacemaker.

Finally, I suggest that approaches and themes deployed by historians of technology to convey to their readers an understanding of machines in general can be used by historians of medicine and technology to discuss the development of medical machines specifically. They, like machines in general, are invented, developed, and subject to innovation, which brings them to the market. Governments, businesses, universities and other organizations shape evolving medical machines, too. Users can shape medical machines, even though this has been a rare event. We should also place medical machines in a cultural context, as we do machines in general. In short, historians of technology can move into the history of medicine field, especially when the subject is medical machines. If I were younger, I would make this move.

Notes

1 A.S. Relman, 'The New Medical-Industrial Complex,' *New England Journal of Medicine*, 303 (1980), 963–70.
2 D. Dranove, *The Economic Evolution of American Health Care: From Marcus Welby to managed care* (Princeton, N.J.: Princeton University Press, 2000), p. 45.
3 T. Khanna and J. Weber, 'General Electric Medical Systems, 2003,' case study, Harvard Business School, published January 30, 2002, revised October 27, 2005.
4 T.P. Hughes, 'Networks of Medical Power,' unpublished essay prepared for the Stanford Medicine and Humanities Conference, 14 May 2001.

Acknowledgments

Most chapters in this volume are based on papers first presented during a conference at the University of Manchester in July 2003. The editors would like to thank the Economic and Social Research Council, the Society for Social History of Medicine, the Wellcome Trust and the Centre for the History of Science, Technology and Medicine at the University of Manchester for their generous financial support. Thanks also to the colleagues and graduate students at the Centre, as well as the staff of Allen Hall, who all helped to ensure that the conference ran smoothly and successfully.

Notes on Contributors

Julie Anderson is a Research Associate in the Centre for the History of Science, Technology and Medicine at the University of Manchester. She is writing a book on the history of hip replacement with John Pickstone. She has also published on the history of disability and war and is completing a book entitled *The Soul of a Nation: Rehabilitating Bodies in the Second World War*.

Stuart Blume is Professor of Science Dynamics in the Department of Sociology and Anthropology, University of Amsterdam. His publications include *Insight & Industry: the Dynamics of Technological Change in Medicine* (1992); *Poor Health: Social Inequality before and after the Black Report* (co-edited with Virginia Berridge, 2003); (in Dutch) *Limits to Healing: Science, Technology, and the Deafness of a Child* (2006).

Robert Bud is Principal Curator of Medicine at the Science Museum in London. He holds honorary fellowships in Science and Technology Studies, UCL, and in History, Classics and Archaeology at Birkbeck College, London, and is an Associated Scholar in History and Philosophy of Science, Cambridge. He has recently completed a history of penicillin up to the present day, *Penicillin: Triumph and Tragedy* (2007).

Christopher Crenner is Associate Professor and Chair of the Department of the History and Philosophy of Medicine and Internal Medicine at the University of Kansas School of Medicine. His research examines the history of doctor-patient relationships, medical practice, and diagnostic technologies in the twentieth century. He has recently published a book on *Private Practice: In the Early Twentieth Century Medical Office of Dr Richard Cabot* (2005) and also practices and teaches as a general internist at the University of Kansas Medical Center.

Neil Handley is Curator of the British Optical Association Museum at the College of Optometrists in London. He has published and lectured on the changing role of the optician, the history of vision aids, ophthalmic instruments and the diagnosis of ocular disease. He is currently working to gather an international collection of contact lenses and contact lens accessories.

Flis Henwood is Professor of Social Informatics at the University of Brighton. She has published widely on the gender-technology relationship and, more recently, on the use of information and communications technologies in health care. Books include *Technology and Inequality* (2000), co-edited with Sally Wyatt, Nod Miller and Peter Senker, and *Cyborg Lives? Women's Technobiographies* (2001), co-edited with Helen Kennedy and Nod Miller.

Patrik Hidefjäll finished his PhD in Technology and Social Change in 1997 at the Institute of Tema Research, Linköping University with his book *The Pace of Innovation – Patterns of Innovation in the Cardiac Pacemaker Industry*. He has since worked in the medical device industry at Siemens, Biotronik and Gothia Medical as sales representative, area sales manager, strategic analyst and planner and now as marketing director. His most recent article, 'Cardiac Rhythm Management: The Shape of Things to Come', was published in the January 2006 Issue of *Current Opinion in Cardiology*.

Thomas Hughes is Mellon Professor Emeritus of the History of Science at the University of Pennsylvania and Distinguished Visiting Professor at the Massachusetts Institute of Technology. His most recent books include *Human Built World* (2004); *Rescuing Prometheus* (2000); and *American Genesis* (1990).

Gerald Kutcher is Dean's Professor of the History of Medicine in the Department of History, Binghamton University. He has published numerous scientific articles in radiation medicine. More recently, he has published on the history of cancer therapy and military cold war research and on issues in the history of bioethics. He is currently working on a book on post-World War II human experimentation.

Christophe Lécuyer is Principal Economic Analyst at the University of California. He has published extensively on the history of electronics, scientific instruments, and engineering education. Among his recent publications is *Making Silicon Valley: Innovation and the Growth of High Tech, 1930–1970* (2006).

John V. Pickstone is Research Professor in the Centre for the History of Science, Technology and Medicine (CHSTM), University of Manchester. He is the author of *Ways of Knowing: A New History of Science, Technology and Medicine* (2000), and editor, with Roger Cooter, of the

Companion to Medicine in the Twentieth Century (2003). Current projects include cancer, recent health services, and more work on the long history of STM.

Jonathan Reinarz is a Lecturer at the Centre for the History of Medicine at Birmingham Medical School. He has researched and published on various themes in nineteenth-century economic, social and medical history. He has just completed a history of the Birmingham teaching hospitals between 1779 and 1939, the research for his article largely deriving from this project. He has just commenced a history of medical education in provincial England, c.1820–1948.

Carsten Timmermann is a Wellcome Research Fellow at the Centre for the History of Science, Technology and Medicine, University of Manchester. He has published on constitutional therapy and German medicine in the inter-war period and more recently medical research in post-war Germany and Britain, especially on high blood pressure and cardiovascular disease. Currently he is working on a book on the history of lung cancer.

Peter L. Twohig is a Canada Research Chair and Associate Professor at Saint Mary's University, Halifax, Nova Scotia, Canada. He has recently published *Labour in the Laboratory: Medical Laboratory Workers in the Maritimes* (2005) and has co-edited with Vera Kalitzkus *Making Sense of Health Illness and Disease* (2004), *Interdisciplinary Perspectives on Health Illness and Disease* (2004) and *Bordering Biomedicine* (2006).

Takahiro Ueyama is a Professor in the Faculty of Economics at Sophia University in Tokyo. His research and writing range from the fields of economics and economic history to history of medicine and technology. He is writing about the transformation of American Medicine after World War II and has started creating a database on the commercialization of biomedicine. One of his works available in English is *Medicine and the Market: Professionalism, Commerce and Electrical Devices, 1860–1918*, Stanford University Ph.D. Thesis, 1998.

Sally Wyatt is a Senior Research Fellow with the Virtual Knowledge Studio, Royal Dutch Academy of Arts and Sciences. She is one of the editors, with Flis Henwood, Nod Miller and Peter Senker, of *Technology and In/equality: Questioning the Information Society* (2000). Together with Andrew Webster, she is editor of a new book series, *Health, Technology and Society*.

1

Introduction: Devices, Designs and the History of Technology in Medicine

Carsten Timmermann and Julie Anderson

Medicine has been transformed in the last two centuries, and these transformations were intimately linked to technological developments. If we follow the US Office of Technology Assessment and define medical technologies as 'the drugs, devices, and medical and surgical procedures used in medical care, and the organizational and supportive systems within which such care is provided',[1] then medical technology makes up much of modern medicine. Medical technologies changed diagnostic procedures and treatment regimes, and technical innovations were closely associated with new approaches in medical science and with the rise of what we call biomedicine, the marriage between laboratory science and medical practice.[2] They spawned industries and led to new organizations. This volume seeks to understand and explain a number of the devices and designs that transformed medicine, the politics they embodied, the people who devised and used them, the spaces in which the transformations took place, their impact on medical knowledge and practice, and the consequences for practitioners, patients and society.

The introductions to academic books often start by claiming that the issues they address have been neglected by the scholarly community. But it would be wrong to say that technology in modern medicine is a neglected topic, either in the history or the sociology of health and medicine. The devices used by past medical practitioners, were, along with books, the most collectable expressions of medical change; they have always been relevant to medical antiquarian interests, for example.[3] But social historians of medicine have also been interested in medical technologies, and especially in the politics they embodied. Henry Sigerist, for instance, disillusioned with German and American scientific medicine, famously marvelled over the purposeful organization of the

1

Soviet health system in his book of 1937 that was so to impress the British pioneers of the National Health Service.[4] Over the last four decades, through the social and cultural turns in the history of medicine, the interest in the tools and practices of the trade, their origins, cultural meanings, and social functions has become even stronger.[5]

As in the history of medicine, so in the sociology of health and illness, much recent writing has been concerned with new technologies – diagnostic, therapeutic, reproductive and genetic – often linking with that more traditional concern, the organization of healthcare.[6] For example, as Stefan Timmermans and Marc Berg pointed out in an illuminating overview published on the occasion of the 25th anniversary of *Sociology of Health and Illness*, the journal carried an article on technology in its very first issue in 1979.[7] In this paper Peter Conrad discussed the ways in which technologies turned medicine into an institution of social control.[8] Much of the sociological literature on medical technologies in the 1970s and 1980s took the form of critiques, as Timmermans and Berg point out, some, like Conrad's article in a technological determinist tradition, others in a social essentialist mould, depicting tools and devices 'as blank slates to be interpreted and rendered meaningful by culture'.[9] In the 1990s, scholars increasingly influenced by approaches from Science and Technology Studies (STS) began to document the practices associated with medical technologies, often in minute detail. Other authors, in the innovation studies tradition, have focused on the mechanics and economics of innovations in medicine.[10]

This is not the place for a another detailed review of the literature on the history and sociology of medical technologies, such as provided by Timmermans and Berg, or by Harry Marks and by Jennifer Stanton, or the introductions to the collections of essays on innovation in medicine published respectively by John Pickstone, Ilana Löwy and Jennifer Stanton.[11] 'Innovation' was a fashionable term in the early 1990s, when Pickstone and Löwy edited volumes that in themselves were innovative, as Stanton remarks in the introduction to her more recent book on the subject.[12] Since then a lot more work has been done, linking the history and social studies of medical technology to other traditions (and more closely to one another). Much of the new literature on medicine has emphasized the central role of practices (and thereby implicitly of technologies). Museums that collect medical tools and devices have become more interested to understand and display artefacts as part of a social context, including the perceptions of users.[13] The present depth and extent of this interest are evident to the editors from the collaborative research projects in which they have been involved over recent years, on hip surgery, high blood pressure therapy or cancer, from many international discussions, and not least from the large number of contributors to the conference in Manchester in 2003 where the papers collected in this volume were first presented.

This volume is meant to reflect that international interest, showcasing the main directions in which the field has developed in the last 15 years. From the conference papers, we have selected a number of case studies that contribute to current debates over uses of technology in medicine, in some of the chapters by illuminating the origins of contentions, in others by identifying the constellations in which new technologies have been put to productive use, and in others again by following the consequences for the various users of medical technologies.

Our selections are eclectic in terms of historiographical approaches, for many of the methodological axes which sociologists and historians of medical technologies have ground over recent decades are now stowed away safely – in a methodological tool shed from where scholars borrow fairly freely and pragmatically. For example, there has been much debate, as mentioned above, about technological determinism and social construction. The volume edited by Stanton, for instance, explores the usefulness of 'SCOT' (the Social Constuction of Technology, an acronym coined by Pinch and Bijker) for understanding the history of innovations in health and medicine.[14] In this volume, the only chapter that explicitly discusses constructionist notions is that by Wyatt and Henwood, drawing on this literature pragmatically to explain the complex ways in which women make sense of hormone replacement therapy. Most papers implicitly embrace constructionist models, while at the same time recognizing that the technologies impose constraints on their interpretation and use, and that the results may 'reframe' the medical problem that the technology addressed. Timmermann's chapter, for example, locates the emergence of a drug treatment for hypertension in a specific post-war context dominated by the British Medical Research Council. Contingency plays a central role in this history; but the availability of drug treatments then tended to change the ways in which society viewed high blood pressure. In such ways, the 'solution' shapes the future of the problem – a kind of path dependency which is clear in the accounts of vaccination and cochlear implants in Blume's chapter. The chapter by Bud shows how new antibiotic technologies, combined with the rise of environmentalism and more sceptical attitudes towards technology in the 1960s, together shaped the course and the meanings of epidemics caused by resistant bacteria. As the field has matured, straw men have crumbled and case studies need no longer be situated in debates over determinism and construction. Similarly, debates over the agency assigned to different types of actors (often associated with actor network theory) have calmed down, and the chapters in this volume rarely embrace any particular theory. Nor are the chapters based on one particular type of source; they draw on press reports, professional journals, unpublished correspondence, interviews, and on artefacts themselves. But to find out more about what to expect, let us now turn to the individual contributions to this volume.

Technical innovation and the emerging economies of modern medicine

The chapters in the first part of the book, by Pickstone, Reinarz, Crenner and Twohig deal with the emerging economies of modern medicine and the ways in which technical innovations related to the settings and organization of medical practice since the nineteenth century. Pickstone looks at orthopedics in the North West of England, Reinarz at voluntary hospitals in Birmingham, Crenner at private laboratories in Boston, and Twohig at X-ray departments and clinical laboratories in provincial Canadian hospitals. The ways in which certain locations and sites shaped particular technologies is central to much recent scholarship on medical technology. Different technologies grew out of and changed with particular constellations of expertise, local traditions, financial means and political will. But technologies also travel between spaces and transform them.

Pickstone's chapter explores the possibility of illuminating complex, long-term developments by way of a longitudinal, local study. In a self-consciously 'whiggish' paper (but without the celebratory tone that originally characterized the genre) he presents us with a history of what we might want to call the Lancashire bone industry, following different generations of practitioners from Victorian bone-setters in craft-based family businesses to today's orthopedic surgeons at Wrightington Hospital, a centre of modern hip replacement surgery within the British National Health Service and a key node in the global medical device industry. This is a study of embeddedness, identifying how specific local contexts gave rise to specific forms of innovation culture in medicine. Employing a model that he has developed in his book, *Ways of Knowing*, Pickstone explores for a classical industrial region, the various forms and depictions of orthopaedic 'work' in medicine and in the wider society, including craft, medical science, service rationalization, academic research, product development and industrial collaborations.[15] The importance of personal links between surgical innovators and local engineers is also central to Anderson's paper on surgical innovations in Wrightington.

Reinarz tells similar stories about hospitals in that metropolis of Victorian metal trades, Birmingham. He draws on Schumpeterian economics to understand the voluntary hospital as a specific setting for innovations. This is an experiment: can Victorian charity hospitals be analyzed in economic terms, like a Schumpeterian firm? Reinarz suggests that, yes, they can to some extent, and that recent historians of medical technology, including Howell and Stanton, in writing microhistories of innovations in their social settings may have (probably inadvertently) followed Schumpeterian models.[16] His study looks at the ways in which innovations were implemented in hospitals which were organized and financed in this particular way. What influence did lay donors and subscribers have? Many of the innovations at voluntary hospitals resulted from donations, so the acquisition of a new piece of equipment was often not discussed in the board minutes. Other innovations, for example surgical methods, came with the appointment of certain practitioners. When hospitals became teaching hospitals, another motivation for innovations was added, but also a potential clash between teaching staff and the hospital board over the use of limited resources. As in other settings, the ways in which technical innovation took place in these hospitals reflected the distribution of power and control.

Crenner studies another important setting for medical innovations, the laboratory, but his context is Boston around the turn of the twentieth century. Unlike most work on medical laboratories, Crenner introduces us to *private* office laboratories run by expert practitioners. The laboratories were closely linked to the consultants' private practices and became the focal points of networks of exchange, relying on 'a system of patronage and personal obligation'. After the turn of the century, Crenner argues, this system gave way to more anonymous commercial laboratories. Crenner explores the circulation of matter (mostly urine) and advice in this system, the technologies employed in analysis and shipping, and he follows the changes in the organization of labor around these laboratory technologies.

Twohig, too, is interested in the organization of labor around new medical technologies in settings where specific technical competence was rare and funds were restricted. He looks at the technical workers in laboratories and X-ray departments in provincial Canadian hospitals in the early-twentieth century, characterized by Stephen Barley and Julian Orr as 'the neglected workforce'.[17] Like the previous chapters, Twohig's deals with issues of status and power around new practices. In a set of detailed case studies, Twohig analyzes the background to debates over the appointment of X-ray and laboratory workers in several hospitals, demonstrating that the technicians often enjoyed a great deal more independence (and responsibility) than generally assumed.[18] Often they were nurses, put in charge of both the clinical laboratory and the X-ray machine, and then asked to take on a range of other tasks 'in their spare time'. With specialist training rarely available, Twohig argues, the roles of these workers were fluid and their skills remarkably broad and flexible.

Context, contingency and the life stories of technologies

The chapters in the middle section of the book are perhaps more conventional in that their focus is on specific innovations, devices and designs and their respective contexts. Handley's and Hidefjäll's chapters explore the histories of specific medico-technical objects, the artificial eye and the cardiac pacemaker, Handley writing from the perspective of a curator and Hidefjäll as a sociologist working in the medical device industry. Ueyama's, Timmermann's and Anderson's chapters locate medico-technological innovations in specific clinical and academic contexts, Ueyama writing on the development of a new approach to radiotherapy for cancer at Stanford University Medical School in the context of cold-war science, Timmermann looking at high blood pressure drugs and the role of the Medical Research Council as a booster of such technologies in post-war Britain and the dominions, and Anderson studying the development of clean-air technologies in the operating theatres of a provincial British teaching hospital specializing in orthopedic surgery.

Not surprisingly, the hospital and the laboratory have been the locus for much work on medical technology, but the role of industry has been covered less well in the history of medicine, and so has the impact of business on medical technologies.[19] This is different in sociology (thanks to interest in biotechnology) and in innovation studies.[20] But the term 'industrial space' can be elastic, as the process of development can take place almost anywhere, whether in a large high-tech facility or the workshops of the producers of artificial eyes discussed in Handley's chapter. The role of industry in collaboration with medicine is evident in Anderson's piece on the development of clean air systems in hip replacement surgery. Timmermann's chapter shows that in Britain immediately after World War II the role of the industry was much smaller than we would instinctively assume; public sector research was then correspondingly more important. But later, in all fields of medical technology, the role of industry became difficult to ignore. The close relationship between industry and medical technology is illustrated by Hidefjäll's article on the cardiac pacemaker manufacturer Biotronik in Berlin, and Blume's chapter points us to the role of vaccine makers in shaping vaccination policies. While the companies studied by Anderson and Hidefjäll occupy very different spaces, both in the market and in terms of physical location, both studies provide us with vital information on markets outside America (which is so often the focus for studies of both invention and innovation). Both studies also demonstrate the role of medical device manufacturers with specialized research and development departments, where much of the work is removed, although not completely, from the hospital setting. Part of the problem of researching the industrial sector is gaining access to the material, as everybody who attempts to research the industrial side of the life history of a medical technology will experience. Industry guards its business information, and in the competitive world of medical technology, maintaining secrecy on techniques, materials and products is very important.

Techniques, materials and products are also a central theme of Handley's chapter, which, like Pickstone's in the first part of the book, takes a long view of its subject. Handley, however, starts from the artefacts that he works with in the collection of the Royal College of Optometrists. We would have liked to include more articles derived from artefacts – not least to illustrate the challenge of 'reading' nontextual sources. Working with artefacts, as curators of museums for science and technology never tire of emphasizing, should ensure that historians gain new perspectives on material functions and understand more of the design and production processes. Handley explores what the objects in his collections, along with what we know about their context, tell us about the historical functions of artificial eyes in particular and prostheses more generally, touching on what we might assume to be modern conflicts over the question whether a device should restore the appearance of an intact body or merely replace lost functions. Of course, artificial eyes, as opposed to spectacles, have always been more about appearance than function, about being seen rather than seeing, and only very recently has it become realistic to think about eye prostheses with which one can see. Like Pickstone, Handley tells us about the craft origins of medical objects. He shows how artificial eyes (and along with them the people who made and fitted them) have become medicalized over the last two centuries or so. Hidefjäll's chapter focuses on a device that was medical from the start. He looks at the innovation culture that gave rise to a succession of cardiac pacemakers produced by the Berlin company Biotronik, founded in 1963 by the young physicist Max Schaldach with his friend Otto Franke, an engineer. Hidefjäll examines the ways in which Biotronik, one of the oldest manufacturers of these devices and a company that in the beginning was 'as much an academic as a commercial venture', was able to maintain and in some cases increase, its market share in the face of competition from large American manufacturers. Unlike many other companies in the medical device industry, Biotronik was not taken over by one of these multinationals. Hidefjäll argues that this was largely due to the size and structure of Biotronik and also the role of its founder Max Schaldach, his hands-on approach to research and development and keen interest in the technical details of the company's products. Appointed to a chair in applied physics in 1970, Schaldach maintained close contacts with university scientists, first in Germany (East as well as West) and later in the US. Hidefjäll introduces us to the specific innovation culture in the company, how it related to its management structures, and how it compared to mechanisms of decision-making elsewhere in the industry.

The chapter by Ueyama and Lécuyer deals with another setting that promoted an innovation culture, Stanford University's Medical School. The focus of his chapter is the development after World War II of a new device for the treatment of cancer with X-rays, the medical linear accelerator. The chapter follows the collaboration between the 'medical entrepreneur' and radiologist Henry Kaplan and the physicist Edward Ginzton, director of the Microwave Laboratory at Stanford, working on the development of a large-scale linear accelerator for high energy physics, and the decisions that led to the development of a radiation device that was more user-friendly and powerful than any of the competing designs. This was a time of rapid expansion in the biomedical sciences in the US and such collaborations between medical researchers and physicists, between big science and medicine were typical of the cold war era. But the turn to high technology in US medicine was not without opponents. Ueyama and Lécuyer analyze the resistance that Kaplan and Ginzton had to overcome in what they interpret as a contribution to a fundamental transformation of medicine at Stanford.

Timmermann in his chapter looks at another episode in the history of what we might want to call 'biomedicalization', the career of the ganglion-blocker hexamethonium, one of the first effective drugs used for lowering blood pressure, also in the immediate post-war period. He shows that a problematic drug became 'successful' as part of a standardized treatment package which included intelligent nurse-technicians and educated patients; like earlier programmes for giving insulin to diabetics, the hexamethonium programmes 'managed' the problematic effects. Timmermann's chapter demonstrates the central role of the Medical Research Council (MRC) for biomedical innovation in Britain and the British dominions. MRC officials effectively supervised the transformation of the ganglion blockers from physiological research tools in Cambridge laboratories (both in Massachusetts and England) through to routine drugs, on a path that involved places as remote as New Zealand. What might now be called 'translational research' was part of the mission of the MRC. In this case, private industry, recruited by the MRC for its purposes, played a rather subordinate role.

Collaborations between surgical innovators and industry also play a central role in Anderson's chapter. Following on from Pickstone's work on the Northwest of England, Anderson's paper concentrates on a particular innovation devised by the orthopaedic surgeon John Charnley in order to control the high rates of infection in hip surgery. In tracing the story of infection in the twentieth century, Anderson demonstrates the ways in which surgeons shifted from a reliance on antibiotics and returned to more physical methods of infection prevention. The special problems of long and deep operations involving prostheses revived questions about the sources of infection – in air and from the bodies of the surgical staff. Germ-free operating enclosures were developed in the 1960s by Charnley in association with a company that designed filtration systems for industry. This collaboration was successful; Charnley reported significantly lowered infection rates and the company sold many of these systems all over the world.

Expectations, outcomes and endpoints

Where the chapters in Part II deal predominantly with successful innovations and the conditions for success, the final part of the book strikes a more cautious note. The chapters by Bud, Kutcher, Blume, and Wyatt and Henwood contribute to present debates over medical technologies by discussing some of the unintended consequences of technical innovations, the difficulties of evaluating technology and (in Wyatt and

Henwood's chapter) of deciding whether to undergo or continue medical treatment.

Bud looks at two episodes in the history of bacterial resistance to antibiotics, an unintended consequence of the widespread use of these drugs since the 1940s. The first of these, a worldwide epidemic of hospital acquired infections with an antibiotic resistant strain of the bacterium *Staphylococcus aureus* in the 1950s, caused much concern among public health experts, but hardly any anxiety in the general public, even when during the influenza pandemic of 1957 the death rates for pneumonia caused by staphylococcus increased alarmingly. In contrast, when ten years later, hospitals in Middlesbrough recorded an accumulation of cases of gastro-enteritis caused by an antibiotic-resistant strain of *Escherichia coli*, this local outbreak caused a national crisis and a broad public debate about the (over) use of antibiotics. Bud looks at what changed between the two events and argues that the alarmed response to the 1967 outbreak (that was far less serious than the earlier episode) was an expression of a general change in the public's attitudes towards the power of science and its products during the 1960s; he points to the debates over nuclear testing and the publication of Rachel Carson's book *Silent Spring*. In this context, the use of antibiotics, especially in modern farming had also become a legitimate subject of public debate and political action.

Changed attitudes towards the fruits of science, along with concerns over the costs of modern healthcare, also fuelled interest in methods that promised to provide objective criteria for the evaluation of new medical technologies. Kutcher in his chapter looks at difficulties of achieving closure in controversies over the effects, comparative risks and benefits of therapies. Focusing on trials of chemotherapy in addition to surgery for breast cancer in the US, Kutcher analyzes the failures of consensus, even for the most meticulously organized trials. He looks at a shift of opinion on the value of adjuvant chemotherapy, pointing to the problem of defining adequate endpoints for studies which reflected the expectations of individual patients and local practitioners as well as satisfying the criteria of the trial organizers. (Such endpoints are also the subject of Blume's chapter). Kutcher suggests that we might understand such problems better if we look at modern clinical trials as metrological practice. Metrology is usually associated with the maintenance of standards for quantities like length, weight and time, but in medicine it frequently refers to the calibration of technologies, with standards residing in procedures and protocols. In the USA, modern clinical trials grew out of large-scale cooperative research during World War II; a coordinating center would devise protocol standards, disseminate them to participating institutions, monitor compliance, and process the collected results; in cancer trials the results often reduced to survival rates. But as Kutcher argues, practitioners and patients are concerned with clinical outcomes, which are complex and unique; unlike survival, they are not easily measurable.

Wyatt and Henwood in their chapter unpack some of the expectations of patients and attitudes towards controversial therapies by analyzing how menopausal women in the UK thought about hormone replacement therapy before and after the abandonment in 2002 of a large-scale clinical trial in the US which resulted in a mass-media debate over the risks and benefits of such treatments. Drawing on some of the recent sociological literature on risk, namely work by Lupton[21] which incorporates concepts by Giddens, Beck, Douglas and Foucault, Wyatt and Henwood point to the shortcomings of what they term the technoscientific approach to risk, based on the deficit model of the public understanding of science and technology, the assumption that if people were given the right information by experts and told how to interpret it, their fears and concerns would disappear and their views would resemble those of the experts. Wyatt and Henwood illustrate the complex ways in which the women they interviewed acquire information, negotiate uncertainty and make sense of their own actions, sometimes switching between realist and constructivist notions of risk.

Like the previous two chapters, Blume's contribution deals with the politics of evaluation. Like Kutcher, Blume suggests that standardized evaluation methods such as randomized controlled trials do not yield the universally valid answers their promoters would like them to produce. Looking at two examples, cochlear implants for deaf children and vaccination against polio, he discusses the tensions between, on the one hand, the universal treatment recommendations and the 'rhetoric of evidence', and on the other, the economic interests of manufacturers and the expectations of patients and other users. These tensions find expression in controversies over what constitutes adequate endpoints for evaluation studies. Blume suggests that the consumerist turn in medicine in the last two decades or so has created new obstacles for standardizers, with patient organizations arguing that qualitative data based on patient experience should also have a place in determining the value of new treatments. In the case of cochlear implants, he argues, the mostly technical evaluations of these devices largely ignored the expectations of parents and the meanings that sign language carries in the deaf community. In some cases, Blume suggests, the recommendations cannot be implemented because they simply do not influence people's decisions despite being ethically sound and logically coherent (he uses the notion of an 'empirical slippery slope'). Global recommendations on polio vaccination, he argues, have generally ignored the path dependency of national vaccination policies and their embeddedness in institutional commitments. He concludes that effective technology assessment has to take into account social policies, patient attitudes and national particularities.

Blume's reflections on endpoints and the evaluation of medical technologies by different groups of users provide us also with a suitable endpoint for this volume, in which we encounter a variety of contexts and constellations in medicine, mostly over the last 200 years, where technologies were introduced and used, produced and sold, improved and applied. However, there is a lot that we have left out. Most significantly (like much other work in the history and sociology of modern medicine) we have focused almost exclusively on North America and Western Europe while mostly ignoring non-western, colonial or post-colonial settings.[22] Quite consciously we have also left out some technologies that have been most controversial in recent years, such as reproductive, genetic and biotechnologies.[23] Much has been written about these, and we decided to focus on more mundane, everyday devices and designs.

The chapters in the final part of the book deal with very recent developments in a world that has lost some of its naivety about the benefits of technical progress. Ours is an era of the paradox; we have lost trust in science and medicine but still expect a cure for cancer any day; in Europe we remodel healthcare systems according to consumerist principles, but we still expect the state to provide us with the latest therapies that the pharmaceutical and medical device industries may have on offer; we believe that there should be no price tag on health, but it is clear to us that rationing is reality; we want to live forever but know we can't afford it (and it probably would not be much fun anyway). The papers in this volume are very clearly of this world: some reflect on the obvious benefits of past technical progress, others point to the complex dilemmas created by these very developments. It will be interesting to revisit these sites and themes in, say, another decade's time.

Acknowledgments

The authors wish to thank friends, colleagues and visitors at the Centre for the History of Science, Technology and Medicine, University of

Manchester, for many stimulating discussions and suggestions. Thanks especially to John Pickstone, who helped us greatly in the preparation of this volume.

Notes

1 *Medical Technology and the Costs of the Medicare Program* (Washington, D.C.: US Congress Office of Technology Assessment, OTA-H-227, 1984), p. x.

2 For useful reflections on origins and usage of the term, see P. Keating and A. Cambrosio, *Biomedical Platforms: Realigning the normal and the pathological in late-twentieth century medicine* (Cambridge Mass.: MIT Press, 2003), pp. 49–82.

3 A good example is Henry Wellcome's collection of books and artefacts. See A. Engineer, 'Illustrations from the Wellcome Library: Wellcome and "The Great Past",' *Medical History*, 44 (2000), 389–404.

4 H.E. Sigerist, *Socialised Medicine in the Soviet Union* (London: Victor Gollancz, 1937). On Sigerist, see Elizabeth Fee and Theodore M. Brown (eds), *Making Medical History: The life and times of Henry E. Sigerist* (Baltimore: Johns Hopkins University Press, 1997).

5 F. Huisman and J.H. Warner (eds), *Locating Medical History: The stories and their meanings* (Baltimore: Johns Hopkins University Press, 2004).

6 For some examples of recent work, see M. Lock, A. Young, and A. Cambrosio (eds), *Living and Working with the New Medical Technologies: Intersections of inquiry* (Cambridge: Cambridge University Press, 2000).

7 S. Timmermans and M. Berg, 'The Practice of Medical Technology,' *Sociology of Health and Illness*, 25 (2003), 97–114.

8 P. Conrad, 'Types of Medical Social Control,' *Sociology of Health and Illness*, 1 (1979), 1–11.

9 Timmermans and Berg, 'The practice of medical technology,' p. 101.

10 This type of work has been strongly encouraged in recent years and received much financial support from agencies such as the European Commission or the Medical and the Economic and Social Research Councils in the UK. See the website of the MRC and ESRC Innovative Health Technologies Programme at http://www.york.ac.uk/res/iht/intro-duction.htm (accessed on 23 March 2006). See also A. Webster (ed.), *New Technologies in Health Care: Challenge, change and innovation* (Houndmills: Palgrave, 2006); N. Brown and A. Webster (eds), *New Medical Technologies and Society: Reordering life* (Cambridge: Polity Press, 2004).

11 H.M. Marks, 'Medical Technologies: Social contexts and consequences,' in W. Bynum and R. Porter (eds), *Companion Encyclopedia of the History of Medicine*, Vol. 2 (London: Routledge, 1993), 1592–618. J. Stanton, 'Making Sense of Technologies in Medicine', *Social History of Medicine*, 12 (1999), 437–48; J. Pickstone (ed.), *Medical Innovations in Historical Perspective* (Houndmills: Macmillan, 1992); I. Löwy (ed.), *Medicine and Change: Historical and sociological studies of medical innovation* (Paris: INSERM/John Libbey, 1993); J. Stanton (ed.), *Innovations in Health and Medicine: Diffusion and resistance in the twentieth century* (London: Routledge, 2002). See also R. Schwartz Cowan (ed.), 'Biomedical and Behavioral Technology,' special issue of *Technology and Culture*, 34 (1993), No. 4.

12   Stanton, *Innovations in Health and Medicine*, p. 1.
13   See, for example, G. Lawrence (ed.), *Technologies of Modern Medicine: Proceedings of a seminar held at the Science Museum* (London: Science Museum, 1994). See also two major websites created by the Science Museum, http://www.ingenious.org.uk and http://www.makingthemodern-world.org.uk.
14   A recent volume that discusses the role of users in shaping technologies, not only medical, is N. Oudshoorn and T. Pinch (eds), *How Users Matter: The co-construction of users and technology* (Cambridge Mass.: MIT Press, 2003).
15   J.V. Pickstone, *Ways of Knowing* (Manchester: Manchester University Press, 2000).
16   J. Howell, *Technology in the Hospital: Transforming patient care in the twentieth century* (Baltimore: Johns Hopkins University Press, 1995), Stanton, *Innovations in Health and Medicine*.
17   S.R. Barley and J.E. Orr, *Between Craft and Science: Technical work in US settings* (Ithaca: Cornell University Press, 1997).
18   See also Marks, 'Medical Technologies', pp. 1597–8.
19   A notable exception is work by J.P. Gaudillière and I. Löwy. See for example the collection of essays published in this series, J.P. Gaudillière and I. Löwy (eds), *The Invisible Industrialist: Manufactures and the production of scientific knowledge* (Houndmills: Macmillan, 1998).
20   See note 10.
21   D. Lupton, *The Imperative of Health: Public health and the regulated body* (London: SAGE, 1995).
22   For starting points, see M. Worboys, 'Colonial Medicine' and R.M. Packard, 'Post-Colonial Medicine,' both in R. Cooter and J. Pickstone (eds), *Companion to Medicine in the Twentieth Century* (London: Routledge, 2003), 67–80, 97–112; A. Cunningham and B. Andrews (eds), *Western Medicine as Contested Knowledge* (Manchester: Manchester University Press, 1997); M. Vaughan, *Curing their Ills: Colonial power and African illness* (Cambridge: Polity, 1991).
23   For some examples, see A.R. Saetnan, N. Oudshoorn, and M. Kirejczyk (eds), *Bodies of Technology: women's involvement with reproductive medicine* (Columbus: Ohio State University Press, 2000); R. Davies-Floyd and J. Dumit (eds), *Cyborg Babies: From techno-sex to techno-tots* (New York: Routledge, 1998); J. Edwards *et al* (eds), *Technologies of Procreation: Kinship in the age of assisted conception* (second edition, London: Routledge, 1993); M. Stanworth (ed.), *Reproductive Technologies: Gender, motherhood and medicine* (Cambridge: Polity, 1987).

Part I

Technical Innovation and the Emerging Economies of Modern Medicine

2

Bones in Lancashire: Towards Long-term Contextual Analysis of Medical Technology

John V. Pickstone

Introduction

There are many ways to write the histories of medical technology. As this conference volume shows, the most popular now, at least for short presentations, is the case study – brief biographies of a device or a programme, preferably with a cast that spreads outward from the innovators to the contexts of creation and use. Such studies are often subtle and revealing; they demonstrate context dependency and historical contingency; they illuminate many different genres of innovation and many different sites; they bring together many different species of historian and social scientist. Yet faced with such richness, we might still ask an old question – what might they add up to?

Hackles sometimes rise when such questions are asked. Are we not post-moderns, writing in an age beyond 'grand narratives' and suspicious of analytical categories, lest their imposition distort the 'reality' we do not believe in? So, instead, we refine our cases – and we leave for a rainy day the task of 'meta analysis', of systematizing the factors involved, or of constructing frames in which the cases are compared across space and time. But to my mind, we could now afford to redress the balance; to spend more time on assessing what our case studies can say, in various combinations, in various ways, to various audiences.

There is no reason to suppose that large narratives are inherently more dogmatic and artificial, or less pluralistic and debatable, than any other level of historical analysis or narrative. Case-histories do not come without selection and interpretation – by analysts as well as actors. Nor are macro-histories without meaning for most of the actors: events are open to interpretation on many scales at once, and reflective men and women will have some understanding of the larger historical processes which mould their lives and localities. Micro versus macro does not correspond to fact versus theory. Fuller understandings require historians who can move between the large and small, and between the relatively empirical and the relatively theoretical, recognizing that these axes stand at an angle to each other. One may seek to generalize from a particular case, or seek to situate it in a wider context; and these activities can be profitably interactive. By looking at both similarities and relationships one may be able to build pictures of configurations over time that can be used at various scales.

One possibility is to take sets of 'cases' which are in some sense 'neighbouring', and ask how we might understand them as a collective. The sets could be contemporaneous with each other: perhaps a set of interacting technologies which involve different modes, such as craft and factory production. Economic historians have shown us how to see crafts and small workshops as integral to production systems that also include massive factories with substantial division of process and labour.[1] Though it is clear that modern medicine is similarly heterogeneous in its modes of work, we have few studies which explore this aspect. But if we take such suggestions seriously, we will have to include a time dimension. Modes of work, for example, have different kinds of histories, and many patterns of interaction can only be understood through historical analysis.

In this paper I try out a different angle – taking advantage of the continuities of a related set of medical technologies in a geographical region I know fairly well. The cases I will tell are all about orthopaedics in some general sense, they are all demonstrably connected through family or master-pupil relations, and they all have a common locality – North West England, more specifically the Manchester-Liverpool region over two centuries through to our present. The sequence interests me as a local historian, as a medical historian, and as an analyst of science-technology relations. We know much of the detail reasonably well partly because colleagues of mine have explored the careers and innovations of most of the main actors.[2] Much of this chapter is based on our collective work over the last 20 years, on studies which are published or soon will be – but not in the form of long, analytical narratives. Here I want to explore the possibilities of that form, whilst acknowledging the merits of other approaches.

But even at the beginning of our work, we knew the contours of this long history. We had learned it from the writings or the direct testimony of the medical actors concerned. We could, therefore, now treat this tale as a tribal lore: we could focus on the manner of its transmission and of its extension to new presents, and thus illuminate the changing self consciousness of local professional groupings. In an age of cultural history, where history merges easily with literary analysis, that would no doubt be appreciated – but that is not my purpose here. Instead I take their tale as a draft of history, and I use the results of historians' investigations to construct a sequence of cases or cameos of orthopaedics as practised in the changing contexts of one region over time.

I hope thereby to explore a peculiarly creative strand in the western history of orthopaedics – from 'folk medicine' to global technologies. I explore it as a component in the history of Britain – or more accurately, as we shall see, of the Anglo-American world in which our story was played out. I am particularly interested in the changing forms of work and organization through which the history was made. The tale extends from bone-setters in the industrial villages and towns of early-nineteenth century Lancashire, to the multinational companies which presently make orthopaedic prostheses. All my sites of practice lie, in some sense, between medicine and industry, and my 'thick descriptions' of each of the sites will characterize them through the social history of knowledge-practice relations and in the context of contemporary industrial and social organization.

Some of the analytical frame is drawn from my work on *Ways of Knowing* where I tried to model technological development in terms of craft, rationalized production, and systematic invention, which I related to natural history, analysis and experimentation as ways of knowing.[3] I used the fluid term 'technoscience' for the close and systematic interactions of 'scientific' and 'technical' organizations, such as universities and industrial companies, especially from the late nineteenth century. I included the government involvement which has characteristically been associated with the development of technoscientific products such as pharmaceuticals.[4] But I was also at pains to stress that all these activities had social and moral meanings. My story here begins with domestic industry; it ends with the medical-industrial complex, but I would wish to stress that throughout, medicine, more clearly perhaps than other technologies, is to be judged by its effects on individuals and by its meanings for them.

Act one: craft

Let's begin in Whitworth, a manufacturing village in the hills, three miles north of the manufacturing town of Rochdale, in the classic industrial landscape of Lancashire. By the early nineteenth century, the regional economy was dominated by textiles; it centred on Manchester, chief among the 'great towns' of northern England and the main commercial centre of the global trade in cotton-goods. The 'Whitworth doctors' were members of a family called Taylor.[5] They were skilled in bone-setting and manipulation, with a reputation too for cancer cures. They drew customers from across the region, who took away the 'Whitworth red bottle', a medicament they provided for all their patients. Some patients would stay for days or months in houses kept for that purpose. The family had been established as doctors in Whitworth from about 1750. They were well known nationally and internationally by the early-nineteenth century, and portrayed as part of the robust, pain-tolerant, democratic culture which was especially associated with the hand-loom weavers (who dominated local culture from about 1770 to about 1830). The last of the family to practice at Whitworth was regularly qualified, and died in 1876.

From soon after 1800, the Taylors also had a surgery in the regional conurbation, in Oldfield Road, Salford, a mile or two from the centre of Manchester and about 15 miles south west of their home base at Whitworth. The Salford surgery was also a local landmark, perhaps the family-business equivalent of the (new) public charity dispensaries – but more specialized, more commercial and probably more positively appreciated. It became part of the 'lore of the city' which was collected by late-Victorian antiquarians to give colour and chronological depth to their commercial capital. J.T. Slugg, for example, recalled his boyhood visits to Oldfield Road to watch the operations; he described the several skins of leather which hung in the surgery, ready spread with a brown kind of plaster.'[6]

What kind of medical technology was this, and how should we place it in social and economic history and geography? At the time, bone-setting would have been called a craft or an art, but so would much of medicine, especially surgery. Bone-setting was commonly recognized as a special skill that might be transmitted through families; sometimes it was linked with other manipulative practices, such as those of the smith or the farrier. And as that linkage shows, there was nothing specifically urban or industrial about these long traditions. The first of the Taylors at Whitworth was chiefly a farrier, decades before the coming of the factories. Yet the developing urban context probably did matter, especially for the *scale* of the practice and for its contemporary interpretation.

In several ways, the Taylors were symptomatic of the popular medical practices of the nineteenth century plebeian cultures which are best known for northern England and mid-west America. British historians tend to attribute that culture to industrialization, Americans to the frontier; but those readings are not so far apart as might first seem. The plebeian cultures of the northern cities were rooted in the semi-urban, domestic manufacture which preceded the factories, and the districts concerned were 'frontier' both in the local weakness of aristocratic or ecclesiastical control and the strength of bottom-up movements such as primitive Methodism and Friendly Societies. We know that the form of herbalism organized in the USA by Samuel Thompson was brought to Britain by Dr Albert Isiah Coffin, who failed in London but succeeded in the northern industrial regions, from his base in Manchester. Herbalism, like bone-setting, was a traditional rural practice now adapted to urban marketing and to the auto-didactic, anti-professional sentiment which underlay much of the plebeian medicine both in the US and the UK.[7] The industrial villages of eastern Lancashire were drenched in this sensibility – as hostile to orthodox medicine as to established religion. Rochdale was a national centre of homoeopathy in its self-help, anti-establishment form.[8] We know, too, the popularity of mechanical models of the body and its workings, and this may also have helped underpin the manipulation of bones in England as well as the USA.[9]

Here then was a craft practice extended as a successful family business across several sites and several generations, in regions of dense and growing population. The Whitworth doctors were celebrated throughout the region, and do not seem to have been attacked by regular medical practitioners. By the Victorian period, several sons or apprentices sought regular qualifications; some appear to have practised in the old way, some respected the particulars of the tradition within a wider medical repertoire, some became more critical – as we shall see. As medical legislation and professionalization, from 1815 and around mid-century, drew deeper distinctions between educated and uneducated practitioners, bone-setters seem to have appeared less as a cheap brand of surgeon, and more as useful 'para-medics', to borrow a later designation. Unlike the later osteopaths and chiropractors (and homeopaths), they pushed no rival medical cosmology, save perhaps the mechanical imagery which was easily compatible with ordinary medicine.[10]

It is not clear how their methods may have changed over time – bone-setters did not often sell themselves through novelty, or write about their methods, or train others much beyond the family. We know little of the range of their craft methods, or the informal natural histories of patients and conditions by which they operated. (Indeed, even for today these aspects of 'fringe' practitioners are not well explored – for all our medical sociology and anthropology).

Act two: and medicine

Switch now to Liverpool around 1890, to the port through which passed most of the cotton wool that came to Lancashire from America, and most of the manufactured cottons which Lancashire companies traded to the rest of the world. By the huge docks was the surgery of a general practitioner, Huw Owen Thomas, descended from a Welsh family of bone-setters (the Liverpool equivalent of the Taylors). He was regularly qualified and practiced with his father for two years, before they fell out and the son manifested his suspicion of bone-setters. He still specialized in treating the fractures and other injuries which were all too common around the docks, but, unlike his ancestors, he was a man of science, and a secularist and political radical. His own science was not evolution, nor experimental physiology; it focused on rest, restoration and the healing powers of the body. He became famous for the Thomas splint – a wooden frame to immobilize a fractured leg, much used in the Great War of 1914–18. He published on this and other topics, and gained an international reputation, including American followers. So what kind of 'technologist' did he represent? What did it mean in practice that he was a professional, not an adept of a craft? What shifts of practice and meaning were bound up in his break with his family tradition and his new identity as a medical man?[11]

Perhaps here we need to separate his audiences. To the extent that Thomas had a private practice, among rich and poor, he was known for a particular skill which distinguished him from other doctors; at all social levels, direct recruitment of patients may have resembled that of the bone-setters. That he was officially medically qualified, as well as peculiarly skilled, would presumably have been an advantage, perhaps especially among richer patients. In as much as he was known to other medical practitioners, and the beneficiary of their referrals, it was probably through a mix of direct reputation and his presentations and publications. If we may here underline and contextualize the obvious, medical professionals who had been through schools, read texts, and sat examination, were thereby expected to be working from published methods and to publicize their innovations in ways that other professionals might then follow. By the last half of the century, most practitioners were 'school-trained'. Of course, they had lots of tacit skills they had learned by imitation or apprenticeship, and elite surgeons created professional means of circulation by which they could visit with innovators and learn directly. But officially at least, medical and surgical methods were supposed to be open, published, and thus re-creatable; the method was separable from the innovating practitioner. This would apply whether the method was diagnosis and prescription of a medicine, or whether it involved surgical manipulation and operations, perhaps with special instruments or 'medical aids' such as splints.

That was the 'system of innovation' characteristic of medical devices in the nineteenth and early-twentieth centuries: manufacturers usually followed doctors' designs, putting them into catalogues and trade advertisements. Other doctors learned of them through publications and medical meetings, perhaps sometimes through demonstrations at meetings. Sometimes, manufacturers might take the initiative and develop a new form, perhaps persuading a doctor to endorse it, but there was little organization above the level of the individual practice and the small manufacturing firms.

What then was the relation of Thomas' methods and skills to contemporary medical knowledge? New medical science was not 'applied'; nor it seems was technique guided by any new physiology or pathology. It seems to have been largely empirical, linked with tradition and grounded in belief in the healing power of the body. Fractures were immobilized to allow the body to repair itself. Bones (like the nervous system) would heal if they were not agitated. This 'cosmology' was professionally presented, but was no more precise in its extensions to practice then the 'principles' of most crafts. Yet it shaped the practice and its image. Fracture care was linked into an older tradition of 'orthopaedics' as bodily education – as the art of making children straight. Patients spent months in splints and special beds. The aim was not just mechanical mending, but restoration of physiological function. Patience, with massage and exercise, brought nature's own rewards.[12]

More technical medical science, it seems, was chiefly important in deciding when and how particular methods should be applied; for example, in suggesting when bone lesions might be damaged by manipulation, rather than ameliorated. Huw Owen Thomas knew about tuberculosis of bone, and about lesions and adhesions. In his opinion, bone-setters knew nothing about diseases, and they were dangerous for that reason. As a doctor with a diploma, he was heir to the growing natural histories of medicine, to the knowledge of cases cumulated in medical texts and journals. He was also heir to the largely nineteenth-century tradition of dissection and pathological analysis, whence came the classifications of lesions which guided his technique. One recalls a characteristic role of 'science' in technology – to indicate where action may be fruitless or worse – sometimes simply by cumulating experience across practitioners, or, as in this case, by also providing analytical frames which underpinned the collected histories and linked explanations to prescriptions.

Act three: and rationalization

We stay first in Liverpool, by the docks, but switch focus to Thomas' nephew – Robert Jones. Where Thomas was a general practitioner and surgeon who specialized because of family skills and high demand, Jones trained to be a general consultant surgeon when that role was expanding rapidly; but he too found special opportunities in bone and joint surgery. Thomas' career was made around his own dock-side surgery: Jones' likewise, but also around his appointment to a charity hospital. As Roger Cooter suggests, Jones came to combine several elements into a new brand of orthopaedic surgery of which he was the chief protagonist – the family tradition of manipulation, aseptic operative technique, early and extensive use of x-rays, and 'organization'.[13]

Aseptic surgery we may regard as a new set of craft skills and equipment which combined the 'theory-based' practices of antiseptic practice with the more empirical tradition that stressed cleanliness. But it chose its ancestors from the first camp, and it was generally regarded as an application of bacteriology.[14] X-rays entered medicine by accident, but from a physics laboratory, which gave a cachet to the invention, at least in some quarters. Most doctors were content to leave its use to technicians such as photographers and electricians, but a few who were technically minded developed the method in medicine and made it part of their specialization, usually alongside electrical and other physical therapies. One such was Thurston Holland in Liverpool. Robert Jones saw the advantages for orthopaedics and worked closely with Holland.[15]

Jones was also noted for the efficiency of his practice, and for the organization and delegation to assistants which allowed him to work on large numbers of patients. In some sense, any doctor who treated large numbers of patients might be said to be efficient, and often by delegation; but by the 1890s, organization was a key theme in business and social life. Jones had shown and developed his organizational skills when he set up the accident service for a huge construction project in the 1890s – the digging of the Ship Canal linking Manchester to the

Mersey estuary near Liverpool. Arguably this was Britain's first such casualty service. The same skills were extended in the Great War when Jones arranged to be asked to organize the British orthopaedic services. Such skills were seen as American efficiency, and Jones had close American links – some inherited from his uncle; and American orthopaedic surgeons were important partners in the war time service.[16]

Large corporations, whether commercial or indeed municipal, required organizers; management was emerging as an occupation, especially in America. That American industries had become part of the cutting edge of British manufacture was nowhere more evident than when factories were developed in a novel 'industrial park' around the Manchester end of the Ship Canal – for British Westinghouse, Kellogg's foods, and later for Henry Ford.[17] These companies were modern in arrangement and management, and some were also engaged in research. We can note here that by the 1920s, Metro-Vickers (which had originated as British Westinghouse) was a national leader in electronics research. Large industrial labs, for product development as well as quality control, were then becoming more common in Britain, but were still often seen as American. Metro-Vicks collaborated with Manchester University (and Cambridge), especially with the physics departments, as recent research has detailed.[18] This was also the period when 'clinical research' was being advocated in British teaching hospitals, often with American examples and sometimes with American money.[19]

In the late-Edwardian years, Jones' Manchester protégé, Harry Platt, had visited America as a young surgeon. He maintained a life-long interest in the USA and was soon noted for American methods, especially the use of clerical workers for note-keeping at the fracture clinics he developed after the Great War. At Ancoats Hospital, in a working class district of Manchester, he was able to so share the work with other young surgeons so they could each develop their specialisms when that was difficult in major teaching hospitals. Platt ensured that he took charge of the fracture cases and he trained assistants and nurses to use plaster of Paris; by concentrating patients into one facility, the general standard of care could be raised, and postgraduate training could be established for protégés who might in turn find positions as specialists.[20]

So maybe, to this point, we have a three-step model of orthopaedic practice. First, the craft carried by individuals or families. Secondly, and sometimes building on those skills, a professional's promotion of methods which are publicized, with more or less evidence, and applied according to an analytical classification and diagnosis of conditions –

but which still may need direct instruction. And then, additionally, the social organization of services to include several assistants and perhaps several sites. In this third form, orthopaedics was a social technology, a form of corporate organization in the public sector. In medicine as in industry, this was a form of rationalized production, supported by analysis in the form of quality-control and fault-diagnosis.

In some new industries such as electrotechnics and pharmaceuticals, there was also much use of experimental systems; and interactions with university science departments helped produce new knowledge as well as better artefacts. In some other 'industries', and here we may perhaps include orthopaedic surgery with automobile manufacture, laboratory experimentation was limited and the focus was on incrementally improving devices and procedures, plus better organization.

The international spread and national forms of 'organization and efficiency' would be worth studying. My impression is that the English form was predominantly provincial rather than metropolitan. If so, its transatlantic geography may follow that of nineteenth-century individualism – as new liberalism and corporatism was built on and over the classic liberalism of an earlier generation.[21]

But the connections between the new orthopaedics and 'progressive' social organization were deeper than analogy. In the work of Jones, Platt and their peers, orthopaedics was built as part of war-time organization and peace-time welfare – around construction projects (as we have seen) and more generally around accident services for industrial workers and for the victims of the new motor traffic. Children with skeletal weaknesses and especially tuberculosis were crucial to orthopaedists' refashioning of their specialism after the Great War; and TB services were the contemporary model of a wide-ranging, coordinated medical service. As I have argued elsewhere, public medicine in Britain from c.1900 to c.1950 was 'productionist' and 'reproductionist' in its concern with workers and soldiers, mothers and children.[22] Orthopaedics was important for three of these four groups, at least.

That Platt was a major proponent of hospital-service planning and a supporter of the NHS is no accident. Orthopaedics was not a major specialism in the rankings of private medicine; it had been established chiefly through public hospitals. By the 1950s a national hospital service would make room for orthopaedic specialists in all major towns, when in the 1930s they had rarely been able to make a living outside the major regional cities. In such ways, leading doctors promoted public investments which were also professional investments. The public, in as much as we know it through lay leaders, shared those goals. Modern, scientific medicine would enable better services, for all.

If we look at Manchester orthopaedics about 1950, we can see how the dream was coming true. At the teaching hospital, Platt had a large purpose-built orthopaedics department, including his fracture clinic: 'Platt's plaster palace' was a modernist concrete construction, contrasting with the Queen Anne style (c. 1908) of the rest of the Infirmary. Platt then held one of the two British chairs in orthopaedic surgery, and was a major national figure in the politics of academic and consultant medicine. He was a power on the NHS Board which was planning the hospitals of the Manchester region (population around 5 million), and this regional board covered several facilities which were in some sense 'orthopaedic', including convalescent homes for coal miners and a long stay hospital in a spa town (Buxton) which was used for convalescent cotton workers including rheumatic cases. The regional TB service included several sanatoria and a long stay home for children, plus specialist surgical centres including one for bone and joint surgery, at Wrightington near Wigan.[23]

The regional board was headed by Platt's friend from medical school, Sir John Stopford, later Lord Stopford, an anatomist turned educational leader who was also head of the University. Stopford too was well connected with government and major medical charities. Hence Manchester became a major site for research investment by the Nuffield Trust and the Arthritis and Rheumatism Research Council. It was seen as the leading provincial school, and the obvious place to develop clinical specialisms which had not been possible when specialist consultants had to depend on private practice and were not funded for research. Manchester gained chairs or departments in osteo-pathology, rheumatology and occupational medicine. Though American critics may have overestimated the redness of this dawn of 'socialist medicine', post-war Britain was notable for the organization of its newly nationalized industries, including hospitals, and Manchester, like South Wales, witnessed a serious attempt to develop research around the diseases of industrial workers. This, together with Platt's eminence and clout gave a platform for another generation of orthopaedic initiative.

Act four: and research

John Charnley came from a pharmacist's family in an industrial town north of Manchester and a few miles from Whitworth.[24] He qualified at Manchester medical school and became especially interested in orthopaedics during military service in World War II. Back in Manchester he worked his way to a consultant's post, which was now salaried. He undertook extensive research, especially on the effects of compression in causing bone to grow together. He pioneered operations to fix hip joints, reducing pain but also mobility. Some of the work was on animals, but mostly it was clinical, albeit with very substantial analytical contributions from histology and from bio-mechanics. Charnley worked with the professor of osteopathology and learned the histological skills.[25] He was a self-taught engineer who could work with a lathe, and an accomplished artist, and he collaborated with a series of academic engineers to better understand and measure the mechanical properties of the locomotor system. Though initially sceptical about the possibilities of artificial joints, and open fixation of fractures, he got interested in problems of joint lubrication, and the possibilities of low friction plastics. At one stage he almost arranged to devote about half his week to systematic research, drawing on histological and mechanical analyses, plus animal experimentation, to solve practical problems in orthopaedics. But partly for private reasons, he instead set up a programme to develop artificial hips, revising the metal prostheses which were already sometimes used to replace the femur head, and using plastic cups to replace the socket on the pelvis.

To this end, he left the Manchester Royal Infirmary, and set up a department in Wrightington hospital near Wigan (and between Manchester and Liverpool). The hospital had been developed by Lancashire County Council as a centre for treating tuberculosis of bones and joints, which by the mid-1950s was being successfully treated by antibiotics. That Charnley was able to set up a small research and development department depended on a mix of general and local factors: that TB hospitals and orthopaedic surgeons were available (or looking for) other functions, was common across the western world. The AO orthopaedic foundation in Davos, Switzerland, well analyzed by Thomas Schlich, was based in a former sanatorium.[26] That more research money was available after World War II than before was also true of most western countries. That clinicians were free to focus on research rather than private practice was characteristic of post-war Britain where the NHS had made hospital medicine into a salaried profession. That Charnley was able to take over part of a former county council hospital was again a consequence of the nationalization of hospitals under the NHS, the increased power of regional medical elites over 'peripheral' hospitals, and especially Platt's influence on the Manchester Regional Hospital Board. The funding from Action Research was probably also due to Platt, who was a major figure in that charity.

It would be worth checking how much the new state funding for the NHS and universities boosted British clinical research in the informal world rankings. Overall, it seems, America strengthened its world leadership, and visits to America became routine for the would-be elites in several clinical specialities, including orthopaedics. But in some specialisms at least, Britain achieved a certain global presence, and orthopaedics was one of them. Of the surgeons who, before 1945, tried to develop prostheses to replace one or other part of diseased hip joints, the best known were American or French. Of those who tried, post-war, to replace both sides of the joint, most were British.

Charnley had been sceptical of artificial hips, then being developed by McKee at Norwich, but he got interested in the prostheses made in France by Judet. That they squeaked seem to him evidence of excessive friction. As an amateur engineer, he thought he could reduce the torque by reducing the size of the femur head and by using a low-friction plastic, Teflon, for the acetabular component in the pelvis. He began inserting artificial hips in 1958, with good results initially. But the cups wore, the particles caused caseous deposits in the tissues, and the joints failed after two to three years. The technical failure was deeply depressing for Charnley – but less so for some of the patients, who were so pleased with the initial results that they came back to seek replacements.

Charnley had reason to be grateful, but he was not much influenced by patients' views. He liked working in Britain because he was patriotic. Initially, at least, he liked the NHS because it enabled him to get on with his research without worrying about recruiting private patients. British patients were uncomplaining and undemanding – good material on which to perfect an operation. To be sure, patients with untreatable back pains could be a nuisance, but at Wrightington he was safe from them. In America, he knew, the pressure from patients and the competition between doctors was more intense. He was later to become wary of the way they rushed to extend and elaborate his work.

Charnley was rescued from his failure by an accidental industrial contact.[27] His technician received a visit from a salesman of a new plastic then being used for gear wheels in textile looms, but which he thought might be useful in joints. The technician – a local, apprentice-trained engineer – tried it out and eventually convinced his master that it was indeed promising. The new material was used in hips, without publication until Charnley was sure he had results better than any of the rival models. To be sure, without that hard plastic hip-replacements would still have developed – some inventors persisted with metal on metal joints, especially when worries mounted about the effects of plastic micro particles – but the new plastic meant that Charnley's designs dominated British sales and were widely copied in continental Europe and the USA.

What kind of technical system was the Charnley set-up? Like Huw Owen Thomas, he was a surgeon with practical gifts (he had a workshop at home and prepared prostheses there); he was also outspoken and often brusque, without the diplomatic skills of Jones and Platt. He benefited from state-medicine supplemented by charity funds; and he worked with a good practical technician. For the plastic, as well as for his later 'germ-free environments', he drew on links with local industry. His instruments and then his prostheses were made by a firm of instrument makers in Leeds with whom he maintained close relations. We might characterize the configuration as surgical craft operating in conjunction with small craft-based companies, but on the basis of state support for a medical service, for academic research, for mechanical and biological analysis, and for product development.

Charnley respected and developed craft skills, and he carefully documented the natural history of hips, both natural and artificial, by collecting post-mortem material wherever possible. He mobilized various kinds of analysis, and he rationalized and regularized his surgical procedures, publishing them in unusual detail, with his own explanatory diagrams. His programme was set up to systematize invention, but there was little place for experimental systems beyond test rigs of various kinds, and attempts to assess the reactivity of various material in the body. For some kinds of orthopaedics, for example the fixation of open fractures, animal experiments were used,[28] but it was difficult to model the human hip joint, at least in Charnley's facilities. This surgical innovation, like much industrial innovation, is best seen as R&D – systematic invention and refinement, complemented by collections of case histories and case materials, and supported by many kinds of analysis.

Act five: to technoscience

We can end this tale by looking at Wrightington now, and then drawing back to assess the relevant geography, 30 years after Charnley's retirement in 1975. Wrightington remains a national centre for hip surgery; it remains part of the NHS and has managed to maintain substantial autonomy across several NHS reorganizations. The operation which Charnley pioneered soon became common enough to be practised in all teaching hospitals and in most district hospitals, as Charnley had predicted, and thus much of Wrightington's present work is now concerned with unusually difficult cases, or with the 'revision' or replacement operations required by infection or structural failures.

For reasons we discuss elsewhere, many different prosthesis designs are presently produced and sold, but the Charnley original still accounts for much of the British market. The design which has overtaken it in statistical terms in Britain was developed in Exeter by an NHS surgeon working with a University department of engineering; it was essentially a redesign of the Charnley hip to fit a different surgical technique. These two designs, plus the Wrightington version of the Exeter, comprise over 90 percent of the artificial hips now being inserted in Britain.[29] If all the world were the NHS, then Charnley might be seen as still central to the present act of our tale. But in the world of medicine, and especially of medical and pharmaceutical industry, the late-twentieth century was increasingly international and especially American. The Anglo-American dimension of our tale had taken a new turning.

Charnley's prosthesis was rapidly imitated in Europe and America, and it substantially influenced the rival designs which were developing in the 1960s. Technical competition intensified as American 'orthopaedic' firms entered the market and offered variants of the designs and materials. American surgeons generally preferred prostheses made from chrome-cobalt rather than stainless steel, they were suspicious of the use of cement, and some found Charnley's surgical procedure too radical, and so, though metal on plastic designs soon dominated the American market, few of them were made by Charnley's manufacturer.

Until the late-1970s, in the USA as in the UK, the manufacturers were followers rather than leaders, which is not to say they were passive or unimportant. Several of the surgeons concerned were good enough engineers to make experimental joints, and they also worked with practical and/or academic engineers for the refinement and testing of designs. Surgeons had a long tradition of dealing with instrument makers, and orthopaedic surgeons in particular were used to working with the craft-based orthopaedic businesses which made splints, crutches, special beds, wheelchairs or walking frames. We have seen that Charnley's prostheses were made by the Leeds company, Charles Thackray and Sons, which manufactured his special instruments; some of his rivals worked with orthopaedic companies.

The key point here is that from the 1970s, artificial hips in the USA rapidly became an exciting commercial proposition for research and development teams, some of which were independent companies and some of which were components of big 'orthopaedic' companies such as Zimmer (of the frame) and Stryker (of the surgical bed). In some cases, groups of engineers left a big company and specialized in joint replacements, as when Biomet was formed in 1977 out of Zimmer. The UK firms involved with pioneer replacements were bought by American firms who continued to manufacture the local products. The London Splint Company was bought by Howmedica in 1969, which was then bought by the pharmaceutical firm Pfizer, who sold it to Stryker in 1998. Thackray, with whom Charnley had worked, was bought by Depuy in 1990, and Depuy was bought by Johnson and Johnson in 1998. The only major corporate player now based in Britain is Smith and Nephew, which originated as a chemist shop in Victorian Hull and later specialized in the production of dressings; they got into joint replacement by acquiring, in 1986, Richards of Tennessee, which had originated in 1934 by secession from Zimmer.[30]

Because Depuy was founded (in 1895) in the small town of Warsaw, Indiana, and several of the off-shoot companies remained there (including Zimmer and Biomet), the town is now the world-centre of research on joint development. But Warsaw, Indiana is not a clinical centre, and the companies do not operate their own hospitals (though who could rule it out?). The design and lab testing can be done by the firms directly but the clinical research, as for pharmaceuticals, requires collaborating surgeons. Depuy, for example, continue to work closely with Wrightington, at least on the hip models which are popular in Britain and Europe.

Joint replacements are different from pharmaceuticals – trying new hip designs is much more complex than trying new pills, and the dynamics of training are more complex. For hips, the clinical development site is also likely to function as a site for clinical training, especially for the surgeons, and hence as a means of helping secure product loyalty. And yet, there are strong parallels. The development of new drugs is sometimes pioneered by academic researchers, but the majority of new products and almost all the R&D work is now done by companies. So too now with medical technologies. The dynamics have to be understood as primarily industrial, albeit working with and through professionals and in high regulation environments. Wrightington, Lancashire, remains emblematic of clinical research and development circa 2000, but it cannot be fully understood without Warsaw, Indiana. The research and development methods which were pioneered in pharmaceutical and electro-technics around 1900, not least by American firms in Manchester, are now commonplace in medical technology – to such a point that even American surgeons worry about their loss of professional independence.[31]

Coda

Our local story was never just local, for the industrial villages of Lancashire were built on global trade, as were the docks of Liverpool and the canal which by-passed them and took ships to Manchester. The popular medical cultures of industrial Britain and of the USA were actively exchanging from the mid-nineteenth century, and by the end of the century professional medical cultures were well connected, with Britons beginning to visit America and to see there the growth of corporations and the progress of market-led medical specialisms. In some ways, as we noted, the aspirations of medical professionals imitated contemporary notions of industrial organization; they were certainly related to schemes for improving welfare. Through the early-twentieth century, British and American Progressives exchanged ideas for social and medical organization, and some of them came to fruition in mid-twentieth-century Britain.[32] The creation of the NHS has generally been regarded as a social achievement; our tale here would suggest that its record for technical innovation deserves to be better studied.[33]

We have seen that for much of the twentieth century, medical innovation in Britain, and indeed in the USA, depended on charity hospitals (and universities). It was largely undertaken by professionals who were also academics; up to the Second World War, it was better funded in the USA, where academic clinicians could gain salaried posts. In both countries, the post-World War II years saw a massive increase in public investment. In orthopaedics, as we have seen, the research was cheap by later standards, and the returns to public well-being were high. Much research was analytical, but some was directly on product development, as in the Charnley case. It drew on mechanical analyses and histological studies, etc, but it was also connected with local industry – for the hard plastic and for the filters which he later used to keep germs out of the air of his operating chamber. One sees how the levels of knowledge and practice cumulate and interact: Charnleys hard plastic must have come, more or less directly, from teams of synthetic chemists; and the filters were in some sense descendants of Pasteur and bacteriological analysis; but they both appeared, in practice, as features of the 'natural history' of an industrial region.

It was from this inventive, provincial culture that orthopaedic devices became a global commodity, and a mainstay of a new kind of technoscientific company which could develop arrays of high-tech products and sell them to surgeons and to hospitals across the globe. They have research and development facilities which now outstrip those of public hospitals and universities. They are the points at which the varied expertise of surgeons, metallurgists, engineers, designers, accountants and salesmen are drawn together in powerful new configurations, with or without the backing of parent companies in pharmaceuticals. They work through surgeons, who remain key actors in the development of products and services, but not with the unrivalled authority of the mid-twentieth century. The surgeons' world is now heavily shaped by companies and also by governments – not just through funding but through safety controls on medical innovations and trials, and through bodies which assess the efficacy and cost-effectiveness of devices and treatments. And patients too are acquiring new roles as consumers – not only generally, through the importation of market rhetorics and practices into public services, but specifically, in the US, by orthopaedic manufacturers' advertisements 'direct to patients'.

Our story, then, runs from crafts to technoscience, both for general industrial contexts and for orthopaedics; in both fields it runs from individualism through a concern with organization, to a stress on new products, knowledge-bases and R&D, though not forgetting that older forms too are also developed. But whereas for most of this story, industry was the context and the ideological conditioning for parts of medicine, at our end of the story, medico-industrial companies are central to orthopaedic development. More generally, whereas national industrial production was once a major reason for improving health, since World War II and especially over recent decades, the success of new biomedical industries has itself become a major desideratum of national medical policies.

Of course, this story could be told in other 'Acts' and in other ways. It would look different had I said more, about patent medicines in the late-nineteenth century, or about private physiotherapists in the late-twentieth; or if I had included more on Liverpool. As I mentioned initially, this narrative is in part the Whig history of professional orthopaedics now – but it is also, probably, the history of public appreciation of bone-related novelties, a history of the sites to which you might have been pointed had you come to Lancashire with a fracture or a bad hip and enquired about best treatment.

By all means, think of alternative narratives and analyses. My aim here is not summary as closure, but long analysis as a way of raising contrasts and questions. I hope I have shown that long histories, and comparisons across time as well as space, may be of interest both for historians and wider publics. They may serve as a way of enhancing case-studies, existent or prospective, and thus encourage virtuous circles between narratives of different scopes and degrees of generalization. Such wider, longer narratives may also allow us to better place our present and likely futures, and perhaps to better assess the gains and losses they may involve.

Notes

1 H. Braverman, *Labour and Monopoly Capitalism: The degradation of work in the twentieth century* (New York: Monthly Review Press, 1974).

2 See R. Cooter, *Surgery and Society in Peace and War: Orthopaedics and the organization of modern medicine, 1880–1948* (Houndmills: Macmillan, 1993); R. Cooter and J. Pickstone, 'From Dispensary to Hospital: Medicine, community and the workplace in Ancoats 1828–1949,' *Manchester Regional History Review*, 7 (1993), 73–84. The final result of the Manchester ESRC project on the Total Hip Replacement will be a book, but many of the local aspects of the story are summarized in J. Anderson, 'Innovation and Locality: Hip replacement in Manchester and the North West,' *Bulletin of the John Rylands Library of the University of Manchester*, Special number on Manchester Medicine, vol. 87 (2007), pt. 1.

3 J.V. Pickstone, *Ways of Knowing: A new history of science, technology and medicine* (Manchester, University of Manchester Press, 2000; Chicago: University of Chicago Press, 2001).

4 See the special issue of *Perspectives in Science*, vol. 13, no. 2, 2005, edited by Ursula Klein.

5 See the rich local history by J.L. West, *The Taylors of Lancashire, Bonesetters and Doctors 1750–1890*, Printed and published by H. Duffy, Central Printing Works, Harriet Street, Walkden, Worsley, M28 5QA.

6 J.T. Slugg, *Reminiscences of Manchester Fifty Years Ago* (Manchester, J.E. Cornish, 1881), pp. 55–7.

7 U. Miley and J.V Pickstone, 'Medical Botany around 1850: American medicine in industrial Britain,' in R. Cooter (ed.), *Studies in the History of Alternative Medicine* (Houndmills: Macmillan, 1988), 140–54.

8 J.V Pickstone, 'Establishment and Dissent in Nineteenth-Century Medicine: An exploration of some correspondences and connections between religious and medical belief systems in early industrial England,' in W.J. Shiels (ed.), *The Church and Healing*, Studies in Church History, Vol. 19 (Oxford: Basil Blackwell for the Ecclesiastical History Society, 1992), 165–89.

9 K.A. de Ville, *Medical Malpractice in Nineteenth Century America* (London and New York: New York University Press, 1990).

10 W.G. Rothstein, *American Physicians in the Nineteenth Century: From sects to science* (Baltimore: Johns Hopkins University Press, 1972).

11 See Cooter, *Surgery and Society*, pp. 23–8.

12 On the radical tradition of conservative orthopedics, see Cooter, *Surgery and Society*, esp. pp. 21–2.

13 Cooter, *Surgery and Society*, pp. 30–4.

14 For the latest and most authoritative account see M. Worboys, *Spreading Germs: Disease theories and medical practice in Britain, 1865–1900* (Cambridge: Cambridge University Press, 2000), pp. 150–92.

15 Cooter, *Surgery and Society*, p. 36.

16 *Ibid.*, pp. 128–32.

17 D. Farnie, *The Manchester Ship Canal and the Rise of the Port of Manchester* (Manchester: Manchester University Press, 1980).

18 T. Cooper, 'The Early Development of Scientific Research in Industry: the case of Metro Vickers,' *Manchester Regional Historical Review*, special number on Science and Technology in the Manchester Region, vol. 18, in press.

19 J. Austoker and L. Bryder (eds), *Historical Perspectives on the Role of the MRC, 1913–53* (Oxford: Oxford University Press, 1985).

20 S.V.F. Butler, 'Academic Medicine in Manchester: The careers of Geoffrey Jefferson, Harry Platt and John Stopford,' in *Bulletin of the John Rylands Library of the University of Manchester*, vol. 87 (2007), pt. 1.

21 See for example, M. Sanderson, *Universities and British Industry, 1850–1970* (London: Routledge and Keegan Paul, 1972).

22 J.V. Pickstone, 'Production, Community and Consumption: The political economy of twentieth-century medicine,' in Roger Cooter and John Pickstone (eds), *Medicine in the Twentieth Century* (Amsterdam: Harwood Academic Publishers, 2000), 1–20.

23 J.V. Pickstone, *Medicine in Industrial Society. A history of hospital development in Manchester and its region, 1752–1946* (Manchester: Manchester University Press, 1985).

24 W. Waugh, *John Charnley: The man and the hip* (London: Springer, 1990).

25 J. Charnley, *Compression Arthrodiesis* (Edinburgh and London: E&S Livingstone, 1953), which includes a chapter on mechanics by J.A.L. Matheson, and histological observations by S.L. Baker, the local professors respectively of mechanical engineering and rheumatism research.

26 T. Schlich, *Surgery, Society and Industry. A revolution in fracture care, 1950s–1990s* (Houndmills: Palgrave Macmillan, 2002).

27 Anderson 'Innovation and Locality.'

28 Schlich, *Surgery and Society*, pp. 86–109.

29 National Joint Register, First Annual Report, 2004.

30 See D. Miller, 'Orthopaedic Technology During the Second Half of the Twentieth Century,' in Leslie Klenerman (ed.), *The Evolution of Orthopaedic Surgery* (London: Royal Society of Medicine, 2002).

31 A. Sarmiento, 'The Relationship Between Orthopaedics and Industry Must Be Reformed,' *Clinical Orthopaedics & Related Research*, 412 (2003), 38–44.

32 D.T. Rodgers, *Atlantic Crossings: Social politics in a progressive age* (Cambridge, Mass.: Harvard University Press, 1998).

33 J.S. Metcalfe and J.V. Pickstone, 'Replacing Hips and Lenses: Surgery, industry and innovation in post war Britain', in A. Webster, *Innovative Health Technologies: New Perspectives, Challenges and Change* (Houndmills: Palgrave, 2006).

3
Mechanizing Medicine: Medical Innovations and the Birmingham Voluntary Hospitals in the Nineteenth Century

Jonathan Reinarz

Introduction

Although most historians incorporate and attempt to explain some aspect of change or innovation in their work, few discuss such developments in the context of innovation studies. Studies that do exist, not surprisingly, tend to be written by business and economic historians. In the field of medical history, this neglected subject has undoubtedly been played down given a shift away from the discipline's heroic past, with its emphasis on pioneers and medical firsts.[1] In the past, much medical history work has documented discoveries, such as the development of vaccination and certain surgical procedures, but subsequent work, such as that of Joel Howell, Jennifer Stanton and many other scholars in this volume, have broken with this tradition by emphasizing the importance of social context to processes of innovation.[2] Largely following, probably inadvertently, an approach advocated by the economist Joseph Schumpeter, who encouraged empirical work on innovation at the micro-level of the firm, most current work tends to take the form of detailed, contextualized case studies, free from the influence of 'super-man' theories.[3] Among other things, such studies have done more than most to ensure the decline of the technological determinist model of innovation, with its assumptions that the development of technology is driven by an autonomous, non-social and internal dynamic. Best represented by the work of Marx, the technological determinist school of thought implies that technological change causes social change, not the reverse.[4] Moreover, while technical change is more often seen to be caused by social factors, fewer recent academic works unproblematically convey the idea that change is progress and certain technologies triumph because they are the best and most efficient.[5] Successful technological innovation involves the construction of durable links tying together individuals and non-human entities, the connections between the two often appearing very arbitrary.[6] Alternatively, as put more succinctly by sociologists of technological change, 'artefacts have politics'.[7] The most valuable studies of innovation have made an attempt to uncover these dynamics. As a result, the view that medical institutions are the passive recipients of information, much like the belief that patients are merely passive users of healthcare systems, is no longer tenable.

As argued by neo-Schumpeterians, local or micro-studies are particularly good at revealing the intricate networks of the decision-making process that guide innovations and allow scholars to make far more reliable generalizations about innovations and their diffusion.[8] Voluntary hospital records can be a particularly useful resource in this respect, as they reveal aspects of this process often in greater detail than those of an historically secretive industrial sector. My own research has provided a number of opportunities to engage with this topic, given that it comprises a study of more than half a dozen teaching hospitals in Birmingham during the nineteenth century. Rather than describe a particular innovation in detail, however, this chapter aims to highlight a number of trends concerning innovation at Birmingham hospitals, as well as some difficulties confronting historians using such medical records for the purposes of innovation studies. If anything, this brief survey should hopefully demonstrate the innovation process to be even more complex than is depicted in recent historical studies of business and industrialization. Furthermore, being a study of English voluntary hospitals in the nineteenth century, it aims to emphasize Blume's assertion that the internal organization of a healthcare system is particularly important to the outcome of the innovation process.[9] Put differently, innovations at English hospitals during an age of charitable medicine often evolved and diffused in very different ways than might have been the case under, for example, a state scheme.

Of course, as should already be apparent, a number of other factors besides the particular organization of an existing healthcare system are worth considering in any study of innovation, including the social networks of medical practitioners. For example, medical education in the case of Birmingham's teaching hospitals clearly conditioned the uptake of innovations. As continues to be the case in most medical training programmes today, most students learn to use only those techniques or technologies they encounter in lectures and ward rounds. It is equally important not to downplay politics. If the history of the medical profession in recent years has imparted any clear lesson it is that practitioners, like the artefacts they daily encounter and employ, have politics. Inevitably, the cost of medical technology is also central to any tale of innovation. While scholars have reminded us that social relations condition the uptake of technologies, we still know very little about the prices and costs of innovation. Studies of technological change, for example, only rarely describe the way in which innovations are mediated through money and the market. The limitations imposed by cost constraints are of central importance to this particular study, as voluntary hospitals, largely reliant on charitable donations, periodically experienced financial difficulties throughout the nineteenth century. As a result, questions of choice or control over decision-making, always central to the innovation process,[10] were at times severely restricted or even entirely removed from the hands of the intended users of technology, the medical practitioners, sometimes as a result of conditions set by hospital donors. As such, the findings of this paper may appear to contrast directly with more recent claims of a historical tendency for medical innovation to be most often instituted by the users of technologies, namely doctors and scientists.[11] During the period covered here, the influence of lay donors is undeniable, as put forth by numerous medical historians, who have long emphasized the degree to which lay supporters of medical charities influenced both patient selection and control in previous centuries.[12] Perhaps it is time we acknowledged their influence over other aspects of healthcare in these years, including the uptake of certain medical technologies and treatments.

Like many other historical studies, this paper employs a very broad definition of medical technology in order to concentrate more specifically on actual processes of innovation.[13] Innovations covered will include both the techniques and technologies championed by hospital medical staff, as well as decisions regarding hospital expansion, organization, as well as fund-raising initiatives. While naturally addressing what might be regarded as radical, or major, innovations, which often require institutional reorganization, it also regularly considers minor, or incremental ones, such as the introduction of a new diet or a particular type of bed.[14] Additionally, this study, like that of Stanton, for example, looks less at the act of invention than the diffusion of medical artefacts.[15] More specifically, it argues that occasionally a very different model of innovation developed at nineteenth-century medical charities compared with that described by historians of late-Georgian and Victorian industry, not to mention post-Second World War healthcare systems.

The proliferation of voluntary hospitals

As any introduction to the history of medicine will emphasize, the eighteenth century witnessed a rapid increase in the number of provincial hospitals in England.[16] Inspired by motives varying from guilt to gratitude, these institutions were funded by voluntary charity. This comprised primarily subscriptions, usually averaging a guinea, but also less regular, one-off donations, such as legacies.[17] Over the nineteenth century, such funding diversified as a result of competition and an increase in the number of medical institutions. Consequently, returns from non-charitable sources, such as property and shares, gradually made up an ever-increasing proportion of hospital income, as did payments from patients and, in ever greater proportions, workplace contributions.[18] Much charity also came with strings attached, ear-marked for certain goods and services. Although hospitals received most of their support in cash, sufficient amounts were not always available when required. At such times, the vast number of more frequently overlooked gifts in kind, including beds and bedding, as well as various foods and fuels, increased in importance.

A period of marked medical innovation, the nineteenth century was also an era of repeat financial crises for English voluntary hospitals. Like other historical works addressing healthcare in these years, this chapter reflects the assumption that financial resources in this period were scarce, often to the extent that not all medical needs could be addressed, neither could all services be offered, leading hospital administrators to ration medical care in a number of ways.[19] Frequently, a creative approach was required in order to effect the continuation, let alone improvements, in services. While the subtle mechanisms of charity, like those of technological innovation, are only now being documented, it is evident from existing historical studies that when committees decided to enlarge or renew hospitals, often before sufficient funds had been raised, furniture and fixtures, including medical equipment, were usually obtained 'through a network of less tangible interactions' and supplemented cash donations.[20] Many hospital administrators often appealed directly to businesses in order to acquire a desired instrument or a specific piece of machinery that was otherwise unaffordable. Other times, social networks did not operate as efficiently. In such circumstances, the goods hospital committees received during reconstruction and expansion could be as varied as their donors' charitable motives.

Birmingham's first voluntary hospitals

To most historians, the case of Birmingham appears well-suited to a study of innovation and technological change. To begin with, the town occupies a privileged place in the history of industrialization, and technological innovation was an important part of these develop-ments. In the eighteenth century, contemporary writers spoke of 'the infinite numbers daily inventing machines for shortening business' in towns such as Birmingham, Sheffield and Wolverhampton.[21] In the words of Dr Johnson, 'The age is running mad after innovation; all business of the world is to be done in a new way'.[22] Toys, buckles, buttons and trinkets in particular made Birmingham a byword for the miracles of the new manufacturing processes.[23] Of the new industrial towns it was the first to reach a population of 50,000.[24] By 1800, Birmingham's reputation as a great industrial centre in particular was undeniable, to which Arthur Young's description of the midland's industrial centre as 'the first manufacturing town in the world' bears testimony.[25]

However, the town's impressive industrial growth also appears unconventional given that it owed little to large-scale technological change. There was in fact no technological revolution in Birmingham during the eighteenth century. Home of the famous toy trade, Birmingham was also to become a byword for the survival of workshop industry in the nineteenth century.[26] At the same time, such tradi-tional craft methods became identified with some of the most innova-tive techniques of manufacturing. Though Watt's patent for the steam engine has linked the town's history with the most momentous of technological changes, steam power was of relatively little importance to Birmingham manufacturers until the 1830s. Indeed, as Eric Hopkins has argued, it is doubtful whether steam power can be considered to have had any direct influence at all upon the growth of the town before 1800.[27] By 1815, there were still only about 40 engines in use in the town. Interestingly, one of these early engines was installed in the town's General Hospital in 1781 and was used to pump water into cis-terns located at the top of the building.[28] Compared to other institu-tions, such as the general hospital at Nottingham, which purchased its

first engine in 1813, Birmingham appears to have been eager to introduce the new technologies pioneered by local industrialists.[29] While
the introduction of such machinery to the hospital at this early date
very likely had much to do with the fact that Matthew Boulton and
James Watt both subscribed to the charity, the views of the institution's
administrators regarding steam power effectively demonstrate that
their decision to introduce an engine to the institution did not imply
an unquestioned endorsement of the technology. In fact, should the
medical staff at the General Hospital have had their way, there would
have been even fewer engines in the town. Besides holding its first
inquiry into the effects of steam engines on the health of patients in
1792, the hospital board repeatedly campaigned against the proliferation of engines in the districts nearest the institution.[30] Though
sharing only tenuous links with the main theme of this paper, this
example should nevertheless illustrate that the diffusion of technology
in Birmingham's hospitals was anything but predictable.

While the town's manufacturers were late to introduce steam power
into their workshops, Birmingham was equally slow to construct its
first voluntary hospital. Originally planned and promoted as a 100-bed
institution by John Ash, a local physician, the General Hospital,
Birmingham opened some 14 years after an organizing committee had
first planned its construction.[31] Like many other provincial institutions, which were scaled down during the planning process, instead of
100 beds, the hospital opened with just 40 beds in 1779, decades after
similar institutions opened in Bristol (1737) and Northampton
(1744).[32] Besides affecting the scale of the project, financial constraints
permitted staff jointly to purchase only a single, substantial piece of
medical equipment, in this case an electro-magnetic apparatus, though
more tools and devices were promised when funds improved.[33] Again,
in their choice of equipment, the institution at Birmingham appears to
have followed a pattern noticeable at other voluntary hospitals.[34] More
items soon became available to staff through other means, if they did
not simply use their own instruments and equipment, and a donation
of beds and an extension in 1792 allowed the hospital eventually to
attain its intended size.[35] The building was again cautiously enlarged in
1826 following improvements in local trade, most furniture and
fittings again being donated in kind.

By this time, given the efforts of enterprising local practitioners, a
number of smaller specialist institutions had also been established,
including an Orthopaedic Hospital (1817), an Eye Hospital (1823), and
an Ear and Throat Infirmary (1843). Most specialist medical charities

grew as slowly as the town's first medical institution. However, in marked contrast to the General Hospital, most usually commenced as dispensaries, medical staff attending the sick and infirm in what were not purpose-built institutions, but often disused shops and hotels. Even when space permitted for an expansion in services, most charities opened with no facilities for in-patients. Less cautious was the growth of the medical school (established in 1828) and the Queen's Hospital (1841), its associated teaching hospital and Birmingham's second general hospital. The rapid growth of both schemes was the result of a particularly generous donation from a single donor, Revd Samuel Warneford, who had been attended for several years free of charge by the medical school's founder, William Sands Cox.[36] However, even the existence of a particularly generous benefactor would not ensure the charities' prosperity beyond a single generation. Eventually, with the death of Warneford, the flow of funding reduced and both institutions were reorganized in the 1860s, having been managed in a very private manner by their founder, Sands Cox.

During the second half of the nineteenth century, additional specialist hospitals were commenced in Birmingham, including the Children's (1862), Women's (1871) and Skin Hospitals (1881). Together, by 1900, the wards of all eight voluntary hospitals, which, if not admitting pupils already, had been recognized as associated teaching hospitals in the last decade of the nineteenth century and contained 800 beds, opened to the first medical students of the newly formed University of Birmingham. As teaching hospitals, these medical charities had attained considerable prestige. In many cases, technology offered an additional source of status and, more than other medical institutions, teaching hospitals were expected to appear up-to-date.[37] For this reason alone, staff at the Birmingham Skin Hospital eagerly obtained Robert Koch's lymph to treat lupus in 1890, before its efficacy was even clinically proven.[38] Had its medical committee not purchased the materials to conduct such therapies, staff believed its standing would have suffered. Consequently, these institutions would continue to canvass vigorously for the financial support required to sustain reputations only heightened by their teaching functions.

Hospital planning and innovation

The question of innovation at Birmingham's voluntary hospitals appears very different from that occurring in its industrial sector from the moment the town's first charity was founded. Long before becoming associated teaching hospitals, the flow of information between these institutions, regardless of region, was particularly fluid, a circumstance that contrasts greatly with models advanced by economic historians, who suggest the details of innovation at British firms were typically kept secret before 1914.[39] Despite often competing for patients, prestige and funding,[40] hospitals throughout the United Kingdom, as revealed by their records, regularly and freely shared technological and organization information with one another. In general Victorian charities did not view themselves as competing businesses.[41] Secrecy in medicine after all 'smacked' of quackery.[42] For example, when discussing the most suitable bed for patients during the construction of the original General Hospital in Birmingham, its committee corresponded with and visited a number of Midland hospitals, including institutions at Worcester, Stafford and Leicester.[43] Only after undertaking such thorough investigations did hospital boards attempt to implement new methods or techniques. Similar inquiries were made of hospitals from London to Aberdeen in order to determine diets, wages, the organization of laundry and kitchen facilities, as well as the ideal space between patients' beds in subsequent years, similar tales being recounted in almost every hospital history.[44] Often such visits and inquiries stimulated further unintended technological transformations, as when staff at the Queen's Hospital travelled to the general hospital in Nottingham in order to inspect the institution's disinfecting oven, only to discover a superior wringing machine in its laundry.[45] As one might expect, this free flow of ideas led to considerable uniformity in the way hospitals developed in England, a fact not always commented upon by hospital historians. No where is this as apparent as in the regulations governing these institutions when first established. For example, the rules drafted for the General Hospital in Birmingham hardly varied from those of existing institutions, whether located in Exeter or Edinburgh.

The pattern repeated itself with the establishment of specialist institutions, whether children's, eye or skin hospitals, in the nineteenth century. Those in Birmingham, for example, based much of their organization on a local dispensary, and on the General Hospital once they commenced taking inpatients.[46] However, there are indications that information did not pass as efficiently between these institutions as it did the first general hospitals. For example, unlike the latter institutions, which by this time could be found in each county, specialist institutions were less common, especially in the early-nineteenth century. Some, such as the Orthopaedic Hospital when it opened in

Birmingham in 1817, were extremely rare outside London. Many such hospitals in their earliest days had but a single medical officer, whose absence would result in the cessation of services. While issuing annual reports that differed very little from other voluntary institutions, the premises and practices of certain specialist hospitals had a tendency to vary when specialisms were only just emerging.

Another factor that hindered uniformity among these institutions was the fact that many were started by practitioners who were unable to obtain hospital appointments at existing institutions.[47] Consequently, though often exerting tremendous effort to establish their specialist hospitals, few of their founders brought much experience of hospital organization to these ventures, besides that which might have been picked up while a student on the wards of a general infirmary. More often resembling private practices than hospitals, these institutions often incorporated the diversity this alone implies. Additionally, a hierarchy appears to have existed among medical institutions which ensured that practitioners, as they gained experience over the duration of a career, usually left their posts at specialist hospitals to take up more prestigious posts at larger general hospitals, not the reverse. As a result, the flow of information and ideas tended to be away from these smaller institutions, few experienced medical practitioners having been appointed to vacant posts at specialist hospitals and thereby introducing new techniques and technologies to these charities. Only in the second half of the century, did links between specialist and general hospitals improve, largely because it had become far more common for practitioners to hold multiple hospital posts. Few developments were as conducive to the cross-fertilization of information and ideas. For example, in 1872, a surgeon at the Children's Hospital in Birmingham was for the first time permitted to retain his post although holding an honorary position at the Eye Hospital.[48] Furthermore, with the establishment of eye, dental and ear departments and clinics at the town's two general hospitals, additional opportunities for practitioners at the town's specialist institutions to hold joint appointments materialized.[49] Though many charities often shared lay governors before this date, from the last quarter of the nineteenth century, staff at all of Birmingham's hospitals appear to have become far more familiar with the organization of each other's institutions. Additionally, unlike firms where members of staff or apprentices were to sign contracts certifying that they would not disclose what they saw or learned during their periods of service, hospital walls were particularly pervious barriers.[50] Evidence of innovation at British hospitals appears not only in their annual reports and minute books, but often in those of other institutions, as well as local newspapers and medical periodicals. Given the transparency in this sector of society, historians might justifiably question whether the individual firm is the most appropriate context for the study of innovation.[51]

Origins and acquisitions

Nevertheless, it is worth highlighting some of the difficulties regularly encountered when undertaking research into innovations at voluntary hospitals. Though much of the earliest work on medical technologies originally addressed the question of origins, it is the inception of many innovations, somewhat surprisingly, that is not always apparent in hospital records. Rather than date several such 'firsts' at Birmingham, this section aims to outline briefly some reasons why the introduction of medical technologies at these and other institutions in these years can be somewhat difficult to date with any certainty. For the purposes of this chapter, these have been reduced to three main points, unified by the notion of charity.

First, as a number of innovations at voluntary hospitals resulted from donations, the acquisition of numerous technologies may not be discussed or deliberated in medical committee minutes as was usual hospital procedure when staff purchased new equipment. As such, many innovations do not appear in expenditure accounts. Instead, many appear in random lists of items medical charities received in kind, usually reproduced in the last pages of annual reports. As a result, the introduction of a new therapy or technology, such as a microscope, at most institutions is heralded alongside the acquisition of biblical prints, bird cages and flower bouquets, if at all. Such was the case at the Children's Hospital in 1882 when, for example, a donation of ice and an ice chest appeared amidst other items, including fruits and flowers; items perhaps less promising of therapeutic revolutions.[52] Occasionally, such donations were also announced in the local press, but other times financial constraints prevented any additional publicity.

While medical historians have emphasized the way in which technologies may have contributed to the success of hospitals by attracting prestigious doctors and scientists,[53] my second point inverts this concept by arguing that it may very well have been the appointment of a practitioner that led to the introduction of some new technologies. Just as medical school records reveal that the possession of valuable teaching aids, such as a particularly complete set of botanical or pathological specimens, assisted practitioners in obtaining sought-after teaching appointments,[54] administrators at Birmingham hospitals enlarged their medical armamentarium with each staff appointment. For example, the Birmingham Children's Hospital obtained the use of a haemocytometer (which facilitated blood counts) in the 1870s in this manner, while the only microscope suitable for bacteriological investigations at the General in the 1880s belonged to its pathologist, whose departure was especially lamented in 1890 because his replacement did not possess a similar instrument; his departure resulted in the additional loss of a celluloid microtome.[55] Like its pathologist, the hospital's dental surgeon was also expected to provide his own instruments.[56] Similarly, the death of a member of staff could occasionally result in a sizeable inheritance. In this way, rather than mark the end of an era or the triumph of new paradigms, as memorably advanced by Max Planck, the death of senior members of staff often reinforced their dated methods, much of their medical equipment and library becoming the property of the hospital and used to teach a subsequent generation of medical practitioners.[57]

Finally, occasionally practitioners provided services to hospitals for a number of years before an institution purchased its own equipment or formally appointed their own qualified officer. This is perhaps typified best in the 'familiar story of the entrepreneurial radiologist' recounted by Jennifer Stanton, among other innovation scholars.[58] Efforts undertaken by a Birmingham radiologist, John Hall Edwards, similarly provided the city's smaller hospitals with the benefit of X-rays before this would otherwise have been economically feasible. In exchange for his services, Hall Edwards was able to augment his radiological experience, as the General Hospital, where he held an appointment, like many other institutions to adopt the technology, undertook only limited numbers of X-rays in the late 1890s.[59] For this reason, the Birmingham Orthopaedic Hospital obtained the use of X-ray technology in 1897, at a time when the institution's expenses exceeded its income and would certainly have prevented the acquisition of any such expensive medical equipment.[60] Had this not been the case, the hospital would not have had access to the new diagnostic technology until the first years of the twentieth century, when finances sufficiently improved.

Innovation and the teaching hospital

As these examples suggest, whether a hospital purchased a particular piece of equipment can in fact be irrelevant to the diffusion of innovations in these years. Far more important to the spread of products and processes was whether an institution was a teaching hospital.[61] This factor alone generally ensured that many students became familiar with a particular technology or technique and took this knowledge with them when leaving Birmingham. Were this not the case, certain technologies at local hospitals would have influenced far fewer numbers of practitioners, usually only those who happened to work at a particular institution. Nevertheless, just because a hospital was an educational institution did not immediately ensure techniques would diffuse widely. Even in Birmingham, staff at different hospitals appear to have had varying degrees of success when it came to disseminating their ideas. Though many examples of this phenomenon exist in hospital records, the point is readily made evident by briefly comparing the experiences of two of the town's most innovative surgeons, Sampson Gamgee, who served at the Queen's Hospital, and Lawson Tait, at the Women's Hospital.

Though having made great advances in the treatment of fractures, Sampson Gamgee is best known for his views concerning the treatment of wounds. The particular innovation with which he has become associated is the use of firm, dry absorbent dressings, that were to be disturbed as infrequently as possible following their application. First outlined in print in 1867, his method of dressing wounds differed markedly from that put forward by other surgeons and, most notably in these years, Joseph Lister. Unlike Lister, Gamgee never saturated his bandages with strong antiseptic lotions, believing these to lessen the body's, or tissue's, own resistance to infection.[62] Unlike non-absorbent dressings, which quickly became soggy and had to be changed daily, Gamgee's dressings could be left in place often for more than a week. Having originally read a report in 1842 published by Dr Mayor of Lausanne, who described the use of cotton wool and gauze for dressing wounds, Gamgee essentially emulated the Swiss physician's methods, though introduced a new material, which combined firmness with maximum absorption, to hospital wards.[63] In 1880, Gamgee described the use of his cotton-wool sandwich in the *Lancet*.[64] From this point on, most biographies of the surgeon claim 'Gamgee tissue' became a household name.[65] However, as much as the diffusion of the new product appears to have been the result of the publicity it received in the pages of the nation's leading medical journals, its success was equally reliant on Gamgee's post at the Queen's Hospital, Birmingham's primary teaching hospital since 1841. Being associated with a teaching hospital regularly guaranteed Gamgee, as it did Lister,[66] both an audience and an opportunity to carefully instruct in the application of his methods.

As this and numerous other examples suggest, the teaching hospital was often the ideal platform from which to successfully launch medical innovations. Alternatively, most of medicine's failures, such as Henry Hill Hickman, whose names remain familiar due only to the meticulous efforts of historians, appear to have operated outside such elevated medical circles. Many of these people are often regarded by their contemporaries as 'medical nobodies', having practiced and experimented alone, usually in rural locations.[67] Most also appear to have been denied their rightful places in history, frequently having been beaten to the post by more enterprising medical entrepreneurs, who occupied more prestigious institutional posts, or merely understood the importance of announcing their discoveries in print or to the appropriate, receptive audience, such as a class of students. However, even those fortunate enough to have occupied scarce and much sought-after hospital teaching posts still encountered difficulties when it can to expounding new principles and practices. A particularly valuable example of the way in which the diffusion of innovative techniques could be hindered by a surgeon's institutional affiliation comes from the life and work of the surgeon Lawson Tait.

Educated in Edinburgh, Robert Lawson Tait first practiced medicine in Yorkshire before coming to Birmingham in 1870. Having removed his first ovarian cyst in 1868, during the next two decades Tait made a name for himself in the field of gynaecology by performing and perfecting similarly hazardous operations while surgeon to Birmingham's Women's Hospital.[68] One of the founders of the British Gynaecological Society (founded in 1884), Tait is best remembered for his operation for the removal of diseased ovaries and other equally invasive surgical techniques. At a time when few surgeons dared open their patients' abdomens, Tait was attempting the riskiest of procedures with rates of success that continue to defy belief. Unlike Lister, whose preference for carbolic and antiseptic means few Birmingham surgeons sought to emulate,[69] Tait was an early practitioner of aseptic methods. In 1887, his achievements even appear to have been recognized locally with his appointment to the Chair of Gynaecology at the medical school in Birmingham. Despite establishing an international reputation and securing a distinguished teaching post, Tait's legacy has in recent years required some rehabilitation from historians, given that his career entered a period of decline in 1892, when he was reporting his best results.

The nature of Tait's legacy clearly appears to have been related to the type of hospital where he held his appointment. From its establishment, staff at Birmingham Women's Hospital had been mindful of the need to honor their teaching commitment and medical students, as well as qualified medical practitioners, were welcome to attend for teaching purposes. However, the hospital also restricted its cases to women suffering from diseases of the pelvic organs.[70] As a result, unlike the town's other teaching hospitals, student numbers were limited to two per clinic or ward round. Moreover, its board noted that the objection of any patient to the presence of spectators at examinations and operations was in all cases to be respected. In general, staff took 'every precaution ... to avoid injury to the most delicate feelings'.[71] When the Women's Hospital began to prioritize paying patients of higher social status in 1873, these restrictions were only implemented more rigorously, though student numbers could also decline for other reasons. For example, in 1878, when a decision was made to move the hospital from the town centre to Sparkhill, on the outskirts of Birmingham, student numbers also noticeably declined given the institution's distance from the medical school. By 1889, however, distance no longer determined attendance, as the managing committee revised the institution's rules, prohibiting students from attending the inpatient facilities of the hospital.[72] Neither were they permitted during outpatient hours. As a result, at a time when Tait's methods were attracting greatest attention, his influence on young surgeons was substantially curtailed by the institution's lay governors.[73] For the remainder of his career, Tait negotiated a fine balance between medical education and patient privacy. A strong advocate of practical knowledge during his years as a student, Tait was rarely able to offer his students direct access to such knowledge on the wards of the Women's Hospital.[74] Faced with few other ways to disseminate his techniques, it comes as little surprise that Tait regarded the writing of textbooks as the best form of advertising his methods.[75] Books, however, were no substitute for the methods most students learned by observing and carrying out themselves on patients. Neither could Tait's publications ever convey the full details of his work, given his reluctance to illustrate them with 'costly lithographs'.[76] For this reason, his appointment as consultant surgeon to a women's hospital in Brooklyn, New York, which allowed him to demonstrate his techniques abroad, was as important to his American legacy as were his publications. As a result, while almost any practitioner holding a teaching post at a Birmingham hospital was ensured of disseminating their techniques, as well as any instruments they developed, among those closest to them, Tait's influence appeared to increase the further one travelled from Birmingham. Despite establishing an international reputation through his articles and textbooks, Tait's hospital work provided him with only limited influence throughout the midlands during his career. Like previous studies of innovation, this particular example suggests power and control is important to understanding why certain innovations spread successfully in certain contexts or not.[77] In the case of the Women's Hospital, each of its surgeons throughout this period was denied the full benefits of their teaching posts given the sensitive nature of the particular cases they treated.

Subsequent lives

Unlike the lives of Birmingham's most noted medical practitioners, those of instruments and techniques, much like their dates of origin, as already demonstrated, are far more difficult to uncover.[78] For example, though it may be possible to identify when an institution purchased a piece of equipment, it may, as Howell has demonstrated in the case of X-rays, have taken some time for these instruments to influence patient care in the way one might expect.[79] As well, just because an institution purchased a piece of equipment does not mean it was always used as intended. As other scholars have argued, many technologies and techniques commenced their lives outside wards. For example, through an 'easy bit of lateral thinking' many medical techniques were first employed in distinctly non-medical applications or ways of knowing.[80] For example, scales were used to weigh food and ensure orders had been adequately filled, analytical techniques carried out to ensure food and drugs were not adulterated and sterilizers used to purify products, such as milk and wine. In the same vein, a hospital's first thermometer might just as easily have been used to help staff brew better beer as measure a patient's or ward's temperature. Given such heavy use by members of staff, not to mention students, much equipment, especially thermometers and microscopes, also had very short lives and, when broken, were not replaced for some time. For this reason, while minute books reveal that the Birmingham Children's Hospital purchased a carbolic spray in 1872, next to nothing concerning the instrument's regular use thereafter is known. The fact that another sprayer was purchased just three years later may have been the result of overuse, but, equally, this might indicate dissatisfaction with the apparatus.[81] Other times, instruments were misplaced or borrowed by staff and used in private practices.[82] Consequently, though often appearing in expense accounts, we still know very little about the ways in which new technologies were incorporated into the daily work of medical institutions, or, as importantly, when they fell out of use.[83]

The links between voluntary hospitals and industry were equally important to the innovation process at Birmingham's premier medical institutions, the two sectors having eventually developed a very symbiotic relationship during this period, though not quite to the extent suggested for the post-World War II era.[84] Originally recording very few subscribers from the business community, nineteenth-century subscribers' lists reveal ever-increasing proportions of corporate support as the century progressed. Hospital collectors frequently targeted businesses when canvassing for donations, informing managers of the exact numbers of their employees treated at medical institutions.[85] Others required less encouragement to donate, already recognizing the important work of hospitals, or even the promotional value of appearing on subscription lists and supplying their wares to medical charities. Though perhaps few industrialists used hospitals to showcase their machinery as well as William Strutt of Derby, many other enterprising industrial firms and figures understood the promotional value of such strategic product placement.[86] Benefactors to the Birmingham hospitals ranged from the international Linde Refrigeration Company to small local businessmen, such as Mr Ball, a Broad Street engineer, whose electromagnetic apparatus was donated to the neighbouring Children's Hospital in 1864.[87] In both cases, a company's particular charitable act would have provided the firm with a certain amount of publicity and, more importantly, have influenced the treatment of patients at an institution.

If not granted outright, many medical charities obtained equipment and services at discounted rates or even free of charge. Unlike businesses that purchased goods and services, however, hospitals that obtained equipment in this fashion could not always rely on their suppliers in an advisory or maintenance capacity to the same extent as ordinary paying customers,[88] leading occasionally to more rapid breakdowns. Additionally, while businesses in these years relied primarily on suppliers and hired consultants for purposes of research and development, hospitals continued to rely primarily on other voluntary institutions in order to manage innovations. As a result, they also often appear far less loyal to their suppliers than do the administrators of modern hospitals described by innovation scholars.[89]

The impact of charity

While many examples of technological change at voluntary hospitals seem merely to reinforce what has emerged from recent case studies in the field of innovation studies, I would argue that it is the charitable side of the equation that appears to set developments in this specific period and sector apart from changes occurring in alternative health-care systems. In contrast to the doctor-led and market-based models of innovation described in many historical studies, nineteenth-century English hospitals appear occasionally to have freed themselves from market constraints, even though this required them to relinquish some control over hospital development and frequently tolerate second best techniques and technologies, if not the odd unintended consequence. When determined to expand hospital services, managing committees were always able to obtain additional beds, though, very importantly, not always the 'ideal type' or that deemed most appropriate to the work of the charity. For example, the Birmingham Orthopaedic Hospital tended to receive adult beds throughout much of the nineteenth century, although staff treated primarily children. As a result, for much of this period, staff doubled-up their young patients despite ever-present fears of infection.[90] Furthermore, the design of beds often interfered with the manipulation of limbs, leading medical staff to modify equipment, primarily by sawing off the ends of beds.[91] Charitable donations at other hospitals did not always measure up to the expectations of medical staff, though perhaps not as plainly.

In much the same way, donations in kind allowed lay subscribers to influence other aspects of hospital treatment, such as diets, which were continually modified as a result of donations, including fruits, potatoes and cheese, introduced less as a result of medical concerns than due to local surpluses. For example, thanks to a particularly generous donation in 1892, Bovril was introduced to the diets of patients at the Queen's Hospital.[92] Being teaching hospitals, however, savings made in one department allowed staff to use hospital funds to make other acquisitions, and, in this way, maintain cutting-edge reputations by purchasing equipment deemed absolutely essential to instruction, treatment or professional identity by medical staff. But such methods could also result in the duplication of services, hinder the coordination of specialist services between hospitals and cost the community greater sums in the long run. For example, although a Dental Hospital was founded in 1858, a particular donation led the Children's Hospital to establish its own dental clinic soon after the charity was established. The first gifts to the Children's Hospital, all of which were listed in the charity's annual report for 1861, included a scale, clocks, a set of ward and bath thermometers and various surgical instruments, including a complete set of dental tools.[93] Surely stimulated by the latter, rather arbitrary gift, in December 1861, the hospital board instructed one of their surgeons, Mr Howkins, to organize a dental department at the institution. Though only four 'teeth cases' were seen by hospital staff in 1861, four alone were treated in January the following year.[94] Little additional dental equipment was purchased in the next decade, but the work of the hospital nevertheless gradually grew more specialized.

Hospital supporters also occasionally exercised control over cash donations in a way that could equally affect innovations and the development of hospital services more generally. For example, many subscribers ear-marked donations for particular uses and technologies, as when the Eye Hospital accepted £5,000 from Louisa Ryland with a stipulation that hospital beds remain at 55 for 5 years.[95] In October 1865, John Cornforth gave 100 guineas to the Queen's Hospital on condition that it should be invested and not appear in the newspapers.[96] The complete reverse was the case in January 1893, when Sarah Stokes and her sister expressed their intention of making a donation to the same institution. Unlike Cornforth, the sisters suggested their donation be used to construct a children's ward, chapel or the extension of a specialist department and on no account simply be invested. Moreover, in contrast to Cornforth's modesty, they also requested a tablet be placed in the hospital recording the circumstances of the gift.[97] Additionally, many of the shares received by all voluntary hospitals in these years were received as legacies, as was much property from which ground rents were collected. Consequently, though often managed by industrialists and successful entrepreneurs, hospital balance sheets cannot be read in the same way as nineteenth-century business records. Whether restricted by prudent financial laws or the whims of benefactors, hospital finances were not entirely in the hands of finance committees or those individuals who were, on an annual basis, appointed to manage available resources.

Of course, many similar requests were resisted, such as a promise for 16 guineas from a donor should hospital governors have included a ward for lunatics in the General Hospital's original design.[98] Unlike the Leicester Royal Infirmary, the General's governing board ignored this request, but financial constraints in subsequent years would encourage staff and managing committees regularly to make other concessions.[99] Similarly, governors of Birmingham's smallest charities were more rarely in a position to refuse the often eccentric requests of wealthy benefactors in these years. In this respect, independence for these institutions only came with the arrival of the National Health Service in 1948.

Conclusion

By way of conclusion, it is perhaps important to highlight that the majority of charitable support received by nineteenth-century voluntary hospitals was donated in the form of cash, not kind. In either case, the implementation of innovations was contingent on networks, interest groups, beliefs, politics, among a host of other factors. This paper has attempted to highlight the way in which innovations at nineteenth-century medical institutions at times proceeded on different lines to developments in industry, as well as healthcare systems in which hospitals were less reliant on charity. Above all, the particular organization of healthcare in this period appears to have affected primarily choice and control over changes at English hospitals in the nineteenth century, while other dynamics at work during the innovation process will appear more familiar to innovation scholars.

As for the relevance of this study to medical innovation today, perhaps the conditions attached to some bequests only appear less visible, despite the fact that similar influence over changes is occasionally still exercised. The globalization of medical charities and the reliance of hospitals in the developing world on donations of medical equipment discarded by the West, on the other hand, may share certain traits with the model depicted here, though research in this area has yet to attract the attention of scholars, most of these innovations being described only in a linear, if not heroic, fashion in the charities' own reports.[100] One of the tasks facing a future generation of historians will be to bury similar linear models of innovation and replace them with more sophisticated ones which embody the numerous interactions and feedback loops encountered during the processes of both innovation and diffusion.

Acknowledgments

The author wishes to thank Martin Gorsky and Marsha Henry for commenting on previous versions of this chapter.

Notes

1 J.C. Burnham, *How the Idea of Profession Changed the Writing of Medical History* (London: Wellcome, 1998), p. 8; J. Howell, *Technology in the Hospital: Transforming patient care in the twentieth century* (Baltimore: Johns Hopkins University Press, 1995), p. 17.

2 Howell, *Technology in the Hospital*; Jennifer Stanton (ed.), *Innovations in Health and Medicine* (London: Routledge, 2002), p. 5, 7; S. Blume, *Insight and Industry: On the dynamics of technological change in medicine* (Cambridge: Mass.: MIT Press, 1992); R.S. Cowan, *A Social History of American Technology* (Oxford: Oxford University Press, 1997), pp. 2–4; D. Arnold, *Science, Technology and Medicine in Colonial India* (Cambridge: Cambridge University Press, 2000), pp. 1–2, 9, 12–14.

3 J. Schumpeter, *Business Cycles: A theoretical, historical and statistical analysis of the capitalist process*, 2 vols (New York: McGraw-Hill, 1939); See also R.R. Svedberg, *Joseph Schumpeter: His life and works* (Oxford: Polity, 1991); E.S. Andersen, *Schumpeter and the Elements of Evolutionary Economics* (London: Pinter, 1994); and C. Freeman, 'The Economics of Technical Change,' *Cambridge Journal of Economics*, 18 (1994), 463–514.

4 In particular, see Chapter 15, 'Machinery and Large-Scale Industry,' in K. Marx, *Capital: A critique of political economy*, Volume 1 (London: Penguin Books, 1990), 492–639.

5 P. Lemonnier, 'Introduction,' in Lemonnier (ed.), *Technological Choices: Transformation in material cultures since the Neolithic* (London: Routledge, 1993), p. 16; J. Stanton, 'Introduction,' *Innovations in Health and Medicine*, p. 1.

6 Lemonnier, 'Introduction,' pp. 16–21.

7 D. MacKenzie, *Knowing Machines: Essays on technical change* (Cambridge Mass.: MIT Press, 1998), p. 14; Lemonnier, 'Introduction,' p. 20.

8 R.R. Nelson and S.G. Winter, 'In Search of Useful Theory of Innovation,' *Research Policy*, 6:1 (1977), 36–76.

9 Blume, *Insight and Industry*, p. 44.

10 Lemonnier, 'Introduction,' pp. 5–6; Blume, *Insight and Industry*, pp. 9–10; Howell, *Technology in the Hospital*, pp. 230–1.

11 E. von Hippel, 'The Dominant Role of Users in the Scientific Instrument Innovation Process,' *Research Policy*, 5 (1976), 212–39, pp. 28–9; Stanton, 'Introduction,' p. 4.

12 B.A. Smith, *The Hospitals, 1800–1948: A study in social administration in England and Wales* (Cambridge, Mass.: Harvard University Press, 1964), p. 10; H. Marland, *Medicine and Society in Wakefield and Huddersfield, 1780–1870* (Cambridge: Cambridge University Press, 1987), pp. 117, 156, 175; K. Waddington, *Charity and the London Hospitals, 1850–1898* (Woodbridge, Suffolk: Royal Historical Society, 2000), pp. 51, 106, 166, 174, 211.

13 Howell, *Technology in the Hospital*, p. 7.

14 C. Freeman, 'Economics of Technical Change,' p. 474.

15 Stanton, 'Introduction,' p. 2.

16 See, for example, R. Porter, *Disease, Medicine and Society in England, 1550–1860* (Cambridge: Cambridge University Press, 1993), pp. 30–2.

17 F.K. Prochaska, 'Philanthropy,' in F.M.L. Thompson (ed.), *The Cambridge Social History of Britain, 1750–1950*, Volume 3 (Cambridge: Cambridge

University Press, 1992), 357–94, pp. 358–9; S. Cavallo, 'Charity, Power and Patronage in Eighteenth-Century Italian Hospitals: the Case of Turin,' in L. Granshaw and R. Porter (eds), *The Hospital in History* (London: Routledge, 1990), 93–122, pp. 107–8. Waddington, *Charity and the London Hospitals*, p. 14.

18 Waddington, *Charity and the London Hospitals*, p. 15; J. Reinarz, 'Charitable Bodies: The funding of Birmingham's voluntary hospitals in the nineteenth century,' in S. Sheard and M. Gorsky (eds), *Financing British Medicine, c.1750–2002* (London: Routledge, forthcoming).

19 Blume, *Insight and Industry*, pp. 4, 6; C. Edwards, 'Age-Based Rationing of Medical Care,' *Continuity and Change*, 14:2 (1999), 227–65, p. 228.

20 C. Rosenberg, *Care of Strangers: The rise of America's hospital system* (Baltimore: Johns Hopkins University Press, 1995), p. 339.

21 T. May, *An Economic and Social History of Britain, 1760–1990* (London: Longman, 1995), p. 33.

22 *Ibid.*

23 P. Langford, *A Polite and Commercial People: England, 1727–1783* (Oxford: Oxford University Press, 1992), pp. 68–9.

24 *Ibid.*, p. 418.

25 Eric Hopkins, *The Rise of the Manufacturing Town: Birmingham and the Industrial Revolution* (Stroud, Gloucestershire: Sutton Publishing, 1998), p. xiii.

26 Langford, *A Polite and Commercial People*, p. 651.

27 Hopkins, *The Rise of the Manufacturing Town*, p. 34.

28 Birmingham Central Library Archives (BCLA), General Hospital, Birmingham, Annual Report, 1781, MS 1921/414.

29 Jacob, Frank H., *A History of the General Hospital near Nottingham Open to the Sick and Lame Poor of any County* (Bristol: John Wright and Sons Ltd, 1951), p. 92.

30 BCLA, General Hospital, Birmingham, Minute Book, 1766–84, HC/GH/1/2/4.

31 J. Reinarz, *The Birth of a Provincial Hospital: The early years of the General Hospital, Birmingham, 1765–1790* (Stratford: The Dugdale Society, 2003), pp. 3–13.

32 G. McLoughlin, *A Short History of the First Liverpool Infirmary, 1749–1824* (London: Phillimore, 1978), p. 18.

33 BCLA, General Hospital, Governors Minutes, 1766–84, HC/GH/1/2/4.

34 Jacob, *A History of the General Hospital near Nottingham*, pp. 62–3. McMenemey, William Henry, *A History of the Worcester Royal Infirmary* (London: Press Alliances, Ltd., 1947), p. 64.

35 Reinarz, *The Birth of a Provincial Hospital*, p. 30.

36 J.T.J. Morrison, *William Sands Cox and the Birmingham Medical School* (Birmingham: Cornish Brothers Ltd., 1926), p. 60.

37 Blume, *Insight and Industry*, p. 11; Waddington, *Charity and the London Hospitals*, pp. 104, 114.

38 BCLA, Birmingham Skin Hospital, Skin Hospital, Medical Committee, 1890–1928, MS 1918.

39 D. Edgerton and S. Horrocks, 'British Industrial Research and Development before 1945,' *Economic History Review*, 47 (1994), 213–38, p. 215;

W.J. Hornix, 'From process to plant: innovation in the early artificial dye industry,' *British Journal for the History of Science* (hereafter *BJHS*), 25 (1992), 65–90, p. 86; E. Homburg, 'The emergence of research laboratories in the dyestuffs industry, 1870–1900,' *BJHS*, 25 (1992), 91–111, p. 93.

40 Waddington, *Charity and the London Hospitals*, p. 121.

41 Shapely, Peter, *Charity and Power in Victorian Manchester* (Manchester: Chetham Society, 2000), p. 21.

42 W. F. Bynum, *Science and the Practice of Medicine in the Nineteenth Century* (Cambridge: Cambridge University Press, 1994), p. 167; P. Elliott, 'The Derbyshire General Infirmary and the Derby Philosophers: The application of industrial architecture and technology to medical institutions in early-nineteenth-century England,' *Medical History* (2000), 65–92, p. 88.

43 Reinarz, *The Birth of a Provincial Hospital*, pp. 10–12.

44 G. Haliburton, *The History of the Newcastle Infirmary* (Newcastle-upon-Tyne: Andrew Reid, 1906), pp. 35, 54; McMenemey, *A History of the Worcester Royal Infirmary*, pp. 55, 104–5, 231; M. Railton and M. Barr, *The Royal Berkshire Hospital, 1839–1989* (Reading: Royal Berkshire Hospital, 1989), p. 16; Elliott, 'The Derbyshire General Infirmary,' p. 89.

45 BCLA, Queen's Hospital, Birmingham, House Committee Minutes, 1874–6, HC/QU/1/2/6.

46 BCLA, Eye Hospital, General Committee Minutes, 1823–1857, MS 1919.

47 L. Granshaw, '"Fame and Fortune by Means of Bricks and Mortar": The medical profession and specialist hospitals in Britain, 1800–1948,' in Granshaw and Porter (eds), *The Hospital in History*, 199–220, p. 200.

48 BCLA, Medical Committee Minutes, Birmingham Children's Hospital, 1869–77, HC/BCH/1/4/2.

49 Birmingham Central Library, Local Studies (hereafter BLCLS), Queen's Hospital, Annual Report, 1874; BCLA, General Hospital, Annual Reports, 1891–2, 1899–1900, GHB 430.

50 See, for example, Hornix, 'From Process to Plant,' p. 82; and J. Reinarz, 'Fit for Management: Apprenticeship and the English brewing industry, 1870–1914,' *Business History*, 43:3 (2001), 33–53, p. 47.

51 D.E.H. Edgerton, 'Science and technology in British business history,' *Business History*, 29 (1987), 84–103, p. 85.

52 BCLA, Birmingham Children's Hospital, Annual Reports, 1880–84, HC/BCH/1/14/4.

53 Blume, *Insight and Industry*, p. 12.

54 J. Reinarz, 'The Age of Museum Medicine: The rise and fall of the medical museum at Birmingham's medical school,' *Social History of Medicine*, 18:3 (2005), 419–37.

55 BCLA, Birmingham Children's Hospital, Medical Committee Minutes, 1877–93, HC/BCH/1/4/3; *Ibid.*, Birmingham General Hospital, Medical Committee Minutes, 1884–92, HC/GHB/71.

56 *Ibid.*, General Hospital, Birmingham, Medical Committee Minutes, 1897–1901, HC/GHB/73.

57 T. Kuhn , *The Structure of Scientific Revolutions* (Chicago: University of Chicago Press, 1996), p. 151.

58 J. Stanton, 'Making Sense of Technologies in Medicine,' *Social History of Medicine*, 12 (1999), 437–48, p. 439; Blume, *Insight and Industry*, pp. 22–3.

59 BCLA, General Hospital, Birmingham, Annual Reports, 1897–1900, HC/GHB 432; Howell, *Technology in the Hospital*, pp. 104–17; Blume, *Insight and Industry*, p. 26.

60 BCLA, Orthopaedic Hospital, Annual Report, 1898, HC/RO/Box 14; M. W. White, *Years of Caring: The Royal Orthopaedic Hospital* (Studley, Warwickshire: Brewin Books, 1997), pp. 30, 33.

61 Blume, *Insight and Industry*, p. 10.

62 J.S. Gamgee, 'On the Treatment of Wounds,' *British Medical Journal*, ii (1867), 561–2; See also J. Lister, 'On a New Method of Treating Compound Fracture, Abscess etc. with Observation on the Conditions of Suppuration,' *Lancet*, i (1867), 357–9.

63 B.T. Davis, 'Joseph Sampson Gamgee and the Introduction of Absorbent Cotton Wool as a Surgical Dressing,' *Queen's Medical Magazine*, 52 (1960), 8–11.

64 J.S. Gamgee, 'Absorbent and Medicated Surgical Dressings,' *Lancet*, i (1880), 127–8.

65 H.M. Kapadia, 'Sampson Gamgee: a great Birmingham surgeon,' *Journal of the Royal Society of Medicine*, 95 (2002), 96–100, p. 99.

66 R.B. Fisher, *Joseph Lister, 1827–1912* (London: Macdonald and Jane's, 1977), pp. 164–6.

67 R. Porter, *The Greatest Benefit to Mankind* (London: Harper Collins, 1997), p. 366.

68 J.A. Shepherd, *Lawson Tait: The rebellious surgeon (1845–1899)* (Lawrence, Kansas: Coronado, 1980).

69 R. D'Arcy Thompson, *The Remarkable Gamgees: A story of achievement* (Edinburgh: Ramsay Head Press, 1974), p. 149.

70 BCLA, Birmingham and Midland Hospital for Women, Medical Board Minutes, 1871–1892, HC/WH/1/5/1.

71 *Ibid.*

72 *Ibid.*

73 *Ibid.*, Birmingham Women's Hospital, Governors Minute Book, HC/WH/1/1/1.

74 Shepherd, *Lawson Tait*, p. 13.

75 Shepherd, pp. 43–4.

76 L. Tait, *Diseases of Women* (New York: William Wood, 1879), p. v.

77 Blume, *Insight and Industry*, p. 36; T. Schlich, 'Degrees of Control: The spread of operative fracture treatment with metal implants,' in Stanton, *Innovations in Health and Medicine*, 106–25, p. 120.

78 Pickstone, *Medical Innovations in Historical Perspective*, p. 1.

79 Howell, *Technology in the Hospital*, pp. 117–60.

80 R.A. Buchanan, *The Power of the Machine*, p. 200.

81 BCLA, Birmingham Children's Hospital, Medical Committee Minutes, 1869–77, HC/BCH/1/4/1.

82 BCLA, General Hospital, Birmingham, Medical Committee Minutes, 1892–1897, HC/GHB/72.

83 H. Marks, 'Medical Technologies: Social contexts and consequences,' in R. Porter and W. Bynum, *Companion Encyclopaedia to the History of Medicine*, Volume 2 (London: Routledge, 1993), 1592–1618, p. 1593.

84 See, for example, Blume, *Insight and Industry*.

85   BCLLS, Eye Hospital, Annual Report, 1882; BCLLS, Queen's Hospital, Birmingham, Annual Report, 1848–9.
86   Elliott, 'The Derbyshire General Infirmary,' pp. 66, 84.
87   BCLA, Birmingham Children's Hospital, Medical Committee Minutes, 1861–68, HC/BCH/1/4/1.
88   Edgerton and Horrocks, 'British Industrial Research and Development before 1945,' p. 228; Blume, *Insight and Industry*, p. 65.
89   Blume, *Insight and Industry*, p. 65.
90   BCLA, Birmingham Orthopaedic Hospital, Annual Report, 1877–8, HC/RO/Box 14.
91   *Ibid.*, Medical Minutes, 1884-1903, HC/RO/Box 9.
92   BCLLS, Queen's Hospital, Birmingham, Annual Report, 1892.
93   BCLA, Children's Hospital, Annual Report, 1861, HC/BCH/1/14/1.
94   *Ibid.* Governors Minute Book, 1861–70, HC/BCH/1/2/1.
95   *Ibid.*
96   *Ibid.*, Queen's Hospital, House Committee Minute Book, 1863–66, HC/QU/1/2/1.
97   BCLA, HC/QU/1/1/5.
98   *Aris's Gazette*, 16 Feb 1765.
99   E.R. Frizelle and J.D. Martin, *The Leicester Royal Infirmary, 1771–1971* (Leicester: Leicester Hospital Management Committee, 1971), pp. 78–95.
100 See, for example, International Medical Corps, press release, 15 April 2003, http://www.imc-la.com/pressroom/PR041502-iraq.asp. Accessed 6 June 2003; *MindaNews*, 26 January 2003; *The Flagship*, 13 February 2003; *Economist*, 24 May 2003; the Ghana Plastic Surgery website, http://www.plasticsurgery-africa.org/donatedequipment.htm. Accessed 28 August 2005; or that of the American Medical Resources Foundation Inc., http://www.amrf.com/. Accessed 28 August 2005.

4

Private Laboratories and Medical Expertise in Boston Circa 1900

Christopher Crenner

Introduction

It is easy to think of the growth of laboratory-based medicine at the end of the nineteenth century as a product of expanding hospital practice. The novel methods for analyzing blood, bacteria and bones with the microscope, the culture plate and the X-ray might be pictured as spreading out into general medical use from the increasingly technical hospital practices of the period. The hospital laboratories, by this account, provided a convenient site to assemble the necessary equipment, to train analytic skills and to test their application to the care of sick patients gathered on the wards. Gradual success in hospital practice thus supported progressively wider use of the laboratory. But this version of events inclines us toward too narrow a view of the rise of laboratory medicine. Although the growth of hospital laboratories is the central element of the story, there was a minor, parallel development in the United States outside the hospital in the gradual replacement of private, office-based laboratories with independent, commercial laboratories. Laboratory medicine was not strictly a novelty of hospital practice in the United States, but simultaneously secured an independent place outside the hospital's walls, especially within established, elite medical communities like Boston. A focus on private laboratories highlights the way in which laboratory methods could support autonomous, medical expertise. This paper gives an account of the use of laboratory analysis in Boston in the late-nineteenth century, following the career of one physician through his heyday in private, domestic medical practice, and placing this one practice in the context of broader changes in medicine at the turn of the century.[1]

This addendum to the history of laboratory medicine helps us to reorient an account of its development. Historians of medicine have noted that some elite Anglo-American physicians of the day urged caution against growing enthusiasm for new laboratory skills and data. Laboratory medicine, with its standardized data and routinized methods, might threaten autonomous professional judgment and individualized, personal medical care. In accounting for the rise of laboratory practice, historians have linked this skepticism among British practitioners to the slow diffusion of laboratory medicine into routine practice.[2] In a survey of the issue, Sturdy and Cooter additionally found that laboratory practices entered medicine in Great Britain largely as an instrument of bureaucratic efficiency, first in public health, and then only gradually in daily medical practice with the institutionalization of mass public healthcare in the early-twentieth century. My study of the development of private laboratory work in Boston confirms Sturdy and Cooter's suggestion that 'it was not inherent in the nature of laboratories themselves that they should have developed in this way.'[3] Laboratory medicine in Boston secured an early place among elite practitioners as a judgment-laden, personalized medical service, not as a standardized mechanism of bureaucratic efficiency. This study of the private medical practice of Dr William Whitworth Gannett illustrates an intriguing and unfamiliar use of laboratory analysis in the late-nineteenth century as an expert consultative service. Many of Gannett's methods were eventually standardized and, we might say, de-skilled. But they found initial use as expert practices and smoothed the path to the development of independent commercial laboratories – where use of standardized laboratory methods developed in parallel, outside the bureaucracies of central hospital and public health laboratories.[4]

Expert laboratory practice

To illustrate the unfamiliar, judgment-laden, laboratory methods of William Gannett, I will start with a difficult moment in his early career. In 1885 he found himself facing a delicate choice. A sizeable part of his private medical practice came from providing urinalysis results to other physicians. On 14 July 1885, a regular patron, Dr Morrill, sent a specimen of urine by mail to Gannett's office for examination. The accompanying note from Dr Morrill stated 'I send another specimen of urine from...,' and he named his patient. I will refer to this patient here as Mrs Potter, to protect her confidentiality. This request was fairly typical for Gannett's work. He often received specimens from local physicians in his office, sometimes by post. Gannett's problem was that two days earlier he had received a specimen of urine from Dr Morrill for a patient identified as Mrs Porter – as I will call her to indicate the similarity of the two actual names. Were these perhaps one and the same person, the earlier Mrs Porter and the later Mrs Potter, confused in hasty, handwritten notes from the busy Dr Morrill? Or had Dr Morrill really meant to indicate, in effect, 'I send another specimen of urine, but this time it is from Mrs Potter, a new patient.' Gannett had a significant stake in the answer to this question, in part because he was not in the business of providing his physician-clients with standardized laboratory data, but of diagnosing patients by laboratory methods.[5]

Gannett's difficulty was that he had already written to Morrill two days earlier reporting on the earlier Mrs Porter's urine, which demonstrated, he reported, 'a chronic nephritis of the interstitial type,' that is to say bad news about Mrs Porter's kidneys. This diagnosis was quite significant to Dr Morrill, since it supported a suspicion that he raised in his initial request, writing to Gannett: 'Please examine [the] spec[imen] for evidence of chronic albuminuria.' Was Gannett now being asked to confirm the same grim diagnosis with 'another' second specimen? Or was the request for 'another' new, unrelated patient?

The questions might have been irrelevant for simple, standardized laboratory analysis. Gannett might merely have written back reporting the exact characteristics of the urine sample and have left it to Dr Morrill to sort out the identity of his own patients. Gannett did provide his clients with detailed information on the chemical and microscopic constituents of urine, as well as providing blood counts, pathological analysis of tissue specimens, and, for the appropriate fee, private autopsies. His urinalysis reports, in fact, carried considerably more information about the constituents of urine than later, more standardized urinalysis would. Gannett reported on levels of urophaein, indican, urea, uric acid, chloride, albumen, sugar and bile in urine, as well as noting microscopic constituents of the urinary sediment, like cells, casts and bacteria. But Gannett went further. For his fee he also offered an expert opinion, acting as a kind of consulting physician on laboratory analysis.

Gannett's reputation as a laboratory consultant was put at risk in this exchange. The earlier Mrs Porter could hardly have a <u>chronic</u> nephritis that disappeared in a matter of days. In addition, interstitial nephritis was a grave matter, and rare enough to make its identification in two successive patients worthy of note – especially given Dr Morrill's stated interest in the diagnosis of a chronic condition. Gannett's patrons expected and supported a reliable laboratory expertise with direct relevance to their practice. Another of Gannett's patrons, the surgeon Charles B. Porter, had recently written about a patient with 'bilateral floating kidney' asking, for example, 'to know the condition of the kidneys with regard to operation.' Gannett reported back on the chemical and microscopic constituents of the sample and offered the reassuring conclusion 'no bar against operating.' Gannett provided medical opinions for his patrons, although he was cautious to acknowledge limitations to what he could discern from the urine alone, reporting cautiously in another case, for example, when his findings were 'not incompatible with a renal calculus but there is not enough to enable one to say that this condition exists.'[6]

Gannett relied on a high level of cooperation from his patrons in shaping and sustaining this role as a laboratory consultant. His solution to the problem of the two similar names illustrates the working of this arrangement. Gannett wrote back immediately to Dr Morrill explaining his uncertainty about the new sample, and declining to put his earlier opinion on Mrs Porter to a test. If the new specimen were also from Mrs Porter, he reasoned 'another specimen is not necessary as the diagnosis is sufficiently clear ... chronic albuminuria from chronic renal disease.' His original diagnosis stood. But he also addressed the alternate possibility. If the new specimen originated 'from another individ[ual] I should like a fresh specimen,' with his emphasis on fresh. He had, without saying so, declined to report on this new, ambiguously identified sample.

Gannett's reply was a reasonable finesse of the dilemma. Dr Morrill accepted it readily, writing back that Porter and Potter were indeed two different people, and indicating that he was sending off a new specimen of urine from the latter, Potter, by train, packed in ice – presumably to insure freshness. He seemed eager to meet the technical needs of his consultant.[7]

Gannett's clients valued his laboratory skills and cooperated in supporting his expert role. As Dr Holbrook explained in a note to Gannett in 1888, 'this patient for whom you made the analysis was sure that she had Bright's disease, and nothing would convince her to the contrary but this analysis by an expert.' Along with this expression of gratitude Dr Holbrook enclosed 'a money order for $5,' a typical fee for Gannett in the time.[8]

Gannett, in turn, structured his laboratory practices to preserve his expert status. He tried to gather information from physicians concerning their patients. Based in part on this information, he offered judgments on the significance of his urinary findings. He promoted urinalysis as producing independent objective findings. But he usually filtered his findings through his judgments about the case being considered. This strategy was not a covert manipulation. Gannett shared his deliberations as an openly collaborative endeavor. In separate correspondence in 1887 the same Dr Morrill wrote on an unrelated case to have Gannett's opinion on the urine. He provided Gannett with details about the condition of his patient in his accompanying letter. Gannett wrote back reporting on the urine and saying that he doubted a calculous pyelitis in the patient, and he posed a hypothetical basis for the view: 'had I no knowledge of the case other than that to be gotten from the urine I should still incline to the idea.'[9] Gannett referred to his urinalysis as an independent source of information. But his need to rely on objective findings alone was only hypothetical. Gannett usually took into account the context provided by his consultants.

In addition, Gannett did not cede much room to his clients for their independent judgments. So, when he reported on the chemical constitutes of urine he provided normative results rather than simple measurements. In reporting the levels of urea or uric acid he typically referred to these as 'slightly diminished,' 'n[orma]l' 'not increased,' or even 'improved.' Techniques for the quantification of urea and uric acid in urine were likely known to Gannett from the first, and he did occasionally quantify his results on urea in the final days of his laboratory career. But he typically reported his specific analytic results in the form of judgments. Urea was 'increased' only relative to what Gannett thought that it should be. Published values for typical levels of urea in urine existed in the time, but these values varied from text to text and were often qualified in complex ways. Gannett never referred to the development or use of standard values, nor did he typically identify which of different competing methods of measurement he used. His clients received little in the way of standardized data on which to base an independent judgment, although they occasionally sought such information. One such physician wrote back upon receiving Gannett's urinalysis report, emphatically stating 'I am anxious to know how much [underline as in original] the urea is diminished.' Gannett reported again simply that it was 'much diminished' and declined to quantify his result. Gannett's qualitative judgments seemed to reflect his chosen analytic methods, rather than a translation of quantitative results. So he wrote in another later case for example that 'the reaction with Fehling's and Lowe's tests gave a brilliant orange sed[iment] with urine of full strength and ... with urine diluted 1 to 100 a slight lemon yellow p[recipita]te was formed.'[10]

Private office laboratories

It is rare to find extensive records like Gannett's to support a detailed examination of early laboratory practice. But evidence for the existence of similar private laboratory services in the United States is widespread, especially in cities that had a sufficient number of private physicians to support it. In 1883, in Philadelphia Dr Judson Daland wrote to the editors of the city's *Medical and Surgical Reporter* to announce that he would furnish to his local colleagues a written report regarding his 'thorough chemical and microscopical urinary examinations' for a 'moderate fee.'[11] In 1905, a decade after Gannett had quit the practice, Dr James Lewis was advertising to colleagues in Gannett's old neighborhood in Boston that he would analyze blood, pus, urine, and sputum for patients for specified fees. In other large cities by the first decade of the twentieth century, the back pages of regional medical journals, like the *Charlotte Medical Journal*, the *St. Louis Clinical Reporter*, the *Cleveland Medical Gazette*, the *St. Louis Medical Era*, and the *Chicago Medical Visitor*, customarily listed one or two physicians promoting their laboratory services for their colleagues. These physicians managed the local market in different ways. Dr M.O. Hoge of Richmond, Virginia directed his potential clients to 'send specimens PREPAID, by mail or express in a well-corked bottle;' while Dr Jones of St Louis advised more discreetly that advice on the transport of specimens and billing would be 'Cheerfully given' on request.[12]

Scattered records of other private laboratory practices support the idea that they functioned much like Gannett's. The expression of measurements in normative terms like 'diminished' or 'not increased' was common in urinalysis forms of the day. So too was the practice of adding an opinion or diagnosis at the conclusion of the report. In 1900 Dr Fred Baker, a private physician in Worcester, Massachusetts wrote to his colleague Dr Wheeler in the smaller, neighboring town of Spencer. Dr Wheeler had sent him a urine specimen for analysis and Dr Baker wrote back to report his results. Like Gannett, Baker phrased his findings in normative terms, noting in part that the urea and the uric acid were 'somewhat diminished relatively' and 'increased relatively.' Relative to what, Baker did not say. Taken in conjunction with other features of the case, these findings inclined Dr Baker to his opinion: 'I should expect in time that the urine would clear up.' Five months later Wheeler sent another specimen, allowing Baker to confirm, in part, his earlier prediction, since as he reported 'the amount of albumen was slightly less than that in the specimen [sent before].'[13] Like Gannett, Baker closed his reports with an expert opinion.

These scattered private urinalysis records suggest further that the strategic character of Gannett's communications was shared by his laboratory peers. Among the ephemera in a collection of personal papers from Dr Albert N. Blodgett is a urinalysis report from 1892 from Dr Edward S. Wood hinting at a complex negotiation over the results. Dr Wood reported on the sample's 'increase' in uric acid, urea and urophaein and other features and concluded with the observation that 'this examination shows an exceedingly concentrated condition of the urine and an irritation of the kidneys. These are the usual conditions with chronic arsenic poisoning.' Clearly, Dr Wood had made a significant finding. He continued the report in a strategic vein: 'No message was left with the specimen as to whether you wished it analyzed for arsenic or not. I should advise that it be not analyzed for another month, since at least two months are required for the complete elimination of a single small dose.'[14] Dr Wood's communication suggests that he left something to the discretion of his client concerning the assessment of a potentially controversial substance like arsenic.

Even as these private laboratory practices entered the hurley-burley of American medicine, however, the technical basis for urinalysis was rapidly changing: the methods were becoming more routine, and the tools more readily available. Gannett acquired special laboratory skills in the 1870s during a year of study in Germany, and on his return imported equipment for examining urinary sediments from Germany. By the 1890s both the tools and techniques were less esoteric. Mail-order catalogues in the United States were offering equipment that promised to reduce chemical analysis to a few, easily managed steps; so that the albuminometer and the ureometer, as their names suggested, almost measured the albumen and urea for you. By 1903 Charles Emerson was bragging in the pages of the *Johns Hopkins Hospital Bulletin* that with the new ureometer 'the time is not distant when the urea chart will be the duty of the trained nurse, as well as the temperature chart, for the successful use of the urea tube depends on skillful manipulation rather than scientific training.' The idea that nurses would carry a urea-tube along with their clinical thermometer suggested that urinalysis was becoming, in a manner of speaking, de-skilled. Such routinization of laboratory work had long been an aspiration of some physicians, and perhaps a fear of others. As early as 1884, an editorial in the *Journal of the American Medical Association* opined that the measurement of the urine albumen with the new Esbach albuminometer was 'simple enough to be performed as a daily duty by any intelligent nurse or even by the patient himself.' The days of expert medical urinalysis seemed to be numbered.[15]

The medical community that supported Gannett's practice was similarly in transition. Gannett had relied on a close circle of elite collegiality to sustain his expertise in laboratory analysis. He depended on local social, and even neighborly, connections, and there was more than a hint of special patronage in his arrangements. Gannett started his practice with the support of Dr George Tarbell, a senior Boston physician, who provided space for Gannett's laboratory in the private medical office at 110 Beacon Street where he maintained his own practice. Tarbell was also one of Gannett's first regular clients for laboratory services. Another early client was Dr Arthur Tracy, who soon replaced Tarbell, providing Gannett space in his private medical office. Dr Tracy might drop off a specimen in the morning with a note asking 'Dear Billy, Will you examine this urine to see if there is any trouble in the kidneys?' Dr Chamberlain stopped by Gannett's office himself to pick up a report, but arrived too early, as Gannett reminded himself in his account book. Dr Elliott, in contrast, took the precaution of leaving a note alerting Gannett of his intention to call for results the next afternoon; while Dr Porter was fortunate enough to catch Gannett in the street one day in May 1889 to ask about a test that he had requested.[16]

These arrangements for getting Gannett started and sustaining his private laboratory work may not have been typical or widely shared. But another feature of Gannett's local support was common in the time. Gannett served as the unpaid pathologist at several of Boston's proprietary hospitals during the 1880s, where he provided free laboratory services for patients on the hospital wards. The senior physicians at the same hospitals also numbered among Gannett's most frequent clients in private practice. The unpaid hospital pathologists in the late-nineteenth century often used this post as a stepping stone into a prestigious position on the regular medical staff. Securing such a regular staff position at a well-known proprietary hospital, in turn, lent local status and encouraged the patronage of wealthier private patients. Gannett's paired roles of providing free laboratory services in the hospital and paid services in the private office to the same prominent physicians may have functioned in a similar, mutually supportive manner. When a senior physician at Massachusetts General Hospital, where Gannett finally gained a prized appointment to the regular medical staff in 1891, wrote to Gannett at his private laboratory to ask for a free urinalysis for a private patient he emphasized in his note that he did not 'ask a favor without <u>any</u> return.'[17] [underline as in the original]

Anonymous commercial laboratories

During Gannett's career this tight-knit world of elite medical collegiality was attenuating. The exchange of fees, favors, samples and expert opinions between Gannett and his patrons relied on a system of patronage and personal obligation that was being rapidly overlaid by new professional norms. Laboratory practices were similarly being transformed. The physicians who advertised their services in the early 1900s in the pages of the *Chicago Medical Visitor* or the *Cleveland Medical Gazette* suggested that they faced a less personally supportive, or even solicitous, market than Gannett had in the 1880s. Such advertisements in fact largely disappeared from the medical journals by the late 1910s, presumably as the local support for these practices faded further. In a related development, by 1896 the post that Gannett held as hospital pathologist at the Massachusetts General Hospital had been completely reconfigured with the appointment of James Homer Wright as a permanent, full-time, salaried director for the hospital laboratories and pathology services. Other urban proprietary hospitals made a similar addition to the staff in the early-twentieth century, as pathology emerged as an independent medical specialty. Gannett meanwhile got out of the private practice of expert urinalysis and into routine private medical practice by the 1890s.[18]

A combination of professional reorganization and technical change gradually made urinalysis, and related laboratory analyzes like blood counts, into non-expert practices. But private laboratories evolved in step with these changes. One physician who successfully weathered the change was Dr Francis Lowell Burnett, who operated a private laboratory in the 1910s in the same neighborhood as Gannett. Burnett was a member of a subsequent generation in Boston's elite medical community. Like Gannett, he was a graduate of Harvard Medical School in 1906. In 1910, Burnett opened a private office providing laboratory services in Gannett's old Back Bay neighborhood. Burnett probably did not provide urinalysis, instead offering the newest laboratory techniques of his day. The Wassermann serum test for syphilis, for example, was a recent, frequently sought laboratory resource available from Burnett. Burnett left no organized manuscript records, but the clinical correspondence preserved incidentally in the private practice records of a colleague permits some insight into the operation of his laboratory. For a Wassermann test in 1911, local physicians sent their patients over to Burnett's private office at 51 Hereford Street. The patient paid Burnett's fee directly; Burnett then drew blood and analyzed it for syphilis. Burnett, like Gannett, reported his results to the referring physician, although he occasionally released a negative result also to the patient, reporting back to the referring physician about this convenience. Burnett's Wassermann forms were also reminiscent of Gannett's old urinalysis reports. Burnett disclosed nothing about his methods or detailed results. He reported the syphilis test as positive or negative, and personally signed his name to the report. He was offering a professionally endorsed diagnosis, rather than anonymous laboratory data. But he soon found that he was facing new competition.[19]

Rapid technical, professional and institutional changes had fostered an alternative form for private laboratory practice. Soon after Burnett opened his private medical office he began listing his services in Boston business directories and in the business listings of the Boston phone book, but initially only under his name as a private medical office. This arrangement was consistent with the personal, expert services that he provided. About the same time, however, alternate arrangements had appeared under the laboratory listings in the business directories. In 1912 the anonymous Physician's Laboratory opened at 384 Boylston Street near Burnett's practice. It was a private medical laboratory, listed without any association with a physician's practice. By 1916, when the similarly anonymous Sias Laboratories opened, Burnett was already ahead of the changes. Whatever advantages these arrangements offered, Burnett felt them too and had followed suit. In 1914 in a separate site at 82 Beacon Street he opened the anonymous 'Clinical Laboratory,' no longer directly identified with his private medical practice. All three of these medical laboratories remained under the direction of individual physicians, and all likely continued to depend on the business of local physicians. But they were no longer identified with the private medical practices of individual physicians. By 1922, the privately owned Massachusetts Clinical Diagnosis laboratory and the Physicians Clinical Laboratory had joined the pack in downtown Boston.[20]

These anonymous commercial laboratories gradually replaced the physician's expert practices with repetitive, standardized methods that generated a growing volume of routine diagnostic data. Here this story rejoins the path mapped by Sturdy and Cooter, as laboratory medicine came into alignment with the values of bureaucratic efficiency. The growth of these private clinical laboratories had a counterpart in the growth of central, hospital laboratories. Laboratory monitoring became an important instrument in the management of ever more patients through an increasingly complex process of diagnosis and treatment. The use of repeated testing to identify and calibrate disease and to track and validate a course of treatment became a standard element of practice in an increasingly professionalized institutional context. The laboratory had found a place at the center of routine medical practice.[21]

It is important to note, however, that technical developments in laboratory analysis were not unidirectional. Some methods followed the path of urinalysis, toward greater standardization and at least partial delegation to less-expert practitioners.[22] Blood cell counts at the Massachusetts General Hospital adhered to this pattern. They first appeared in the hospital medical records in the 1890s as signed reports from individual senior physicians, carrying a formal diagnostic opinion as a standard feature. But these reports gradually became more like urinalysis reports, as the associated practices became anonymous, routinized, and lost their expert character. Unsigned slips of paper recording only the number of blood cells were common in the same medical charts by the 1910s. Over this same period of time, however, pathology reports on tissue specimens became lengthier, more complex and continued to bear the signatures of individual physicians with their diagnostic opinions. Differences in the nature of these techniques mattered, of course. So, judging the cancerous appearance of a tissue sample was a more complex skill than counting cells under a haemocytometer. But as historians who have studied the movement of practices between physicians and nurses have noticed, technical complexity only partially determined what practices physicians standardized and delegated, shared, or held onto as exclusive skills. I have argued similarly elsewhere that the measurement of blood pressure was 're-skilled' and introduced as a physician's expert practice around 1910, just at the moment when a less-skilled version of blood pressure measurement was being considered for introduction to routine nursing use. In this early-twentieth-century setting, standardization in the production of medical information was neither inevitable nor driven solely by technical change.

Laboratory medicine was a flexible enterprise and served different ends in different contexts. In the British context, Sturdy and Cooter found that laboratory work was an easily standardized – and standardizing – process that fit neatly within the managerial transformation in early-twentieth-century British medicine. Perhaps it is unsurprising that in the United States, where a managerial reform of medicine was piecemeal and often frustrated, laboratories took a different course. My examination of private laboratories in Boston directs attention to an alternate lineage in the development of laboratory medicine. Gannett and his laboratory peers beginning in the 1880s provided technical consultative services to elite communities of private physicians. Their reports were judgment-laden, normative and carried signed opinions that sometimes showed evidence of being negotiated with the ordering physicians. Laboratory medicine under this guise fit well with the aspirations of leading American physicians of the period who sought greater authority, cohesiveness and autonomy for their profession. With the use of laboratory medicine established, the transition to standardized techniques in the American context took place in step with cultivation of medical expert practice. It is interesting to note, in closing, how rapid technical change in the early-twentieth century allowed physicians to adjust and sort their practices, preserving exclusive control over certain highly skilled practices while allowing others to be taken up in standardized form by nurses and technicians.

Notes

1 J. Howell, *Technology in the Hospital: Transforming patient care in the early-twentieth century* (Baltimore: Johns Hopkins Press, 1996); M. Sandelowski, *Devices and Desires: Gender, technology, and American nursing* (Chapel Hill, North Carolina: University of North Carolina, 2000).

2 S. Sturdy and R. Cooter, 'Science, Scientific Management, and the Transformation of Medicine in Britain c. 1870–1950,' *History of Science*, 36 (1998), 421–66; M. Worboys, *Spreading Germs: Disease theories and medical practice in Britain, 1865–1900* (Cambridge: Cambridge University Press, 2000), pp. 214–5.

3 Sturdy, 'Scientific Management,' p. 447.

4 William Whitworth Gannett, manuscript notebooks, 'Urine Analysis. Volumes I–IV' and 'Blood Counts [1880–85],' Francis A. Countway Library of Medicine, Boston, Rare Books and Special Collections Department. Portions of volumes I and IV are unpaginated with references given here by entry date.

5 Gannett, 'Volume I,' entries for 11 July 1885 and 13 July 1885, and letters tipped in.

6 Gannett, 'Volume IV,' entry 7 May 1889 and letter tipped in; 'Volume I,' p. 137.

7 Gannett, 'Volume I,' p. 15 and letter tipped in.

8 Gannett, 'Volume III,' p. 161 and letter tipped in.

9 Gannett, 'Volume II,' pp. 211–16 and letter tipped in.

10 Gannett, 'Volume I,' entry 23 July 1885 and letter tipped in; and 'Volume I,' p. 83.

11 J. Daland, 'Letter to the Editors' *Medical and Surgical Reporter,* 48 (1883), 701.

12 Advertisement, *Charlotte Medical Journal,* 19 (1901), xxiv; advertisement, *Clinical Reporter (St. Louis),* 18 (1905), 396. Some library bindings of these journals have omitted the advertising pages, but see for similar examples:

Cleveland Medical Gazette, 17 (1901), 63; *Medical Visitor (Chicago)*, 18 (1902), n.p., advertising section. These journals were consulted in Countway Library, Rare Books.

13 Letters, Fred Baker to E.R. Wheeler, 14 February 1900 and 23 July 1900, in Richard Clarke Cabot Papers, Patient Records, Harvard University Archives, Harvard University, Boston, Volume 1, p. 120.

14 Letter, Edward S. Wood to A.N. Blodgett, 13 April 1892, Papers of Albert N. Blodgett, Countway Library, Rare Books.

15 C.P. Emerson, 'The Accuracy of Certain Clinical Methods,' *Johns Hopkins Hospital Bulletin*, 14 (1903), 9–18, quote pp. 17–18; editorial, 'Esbach's Method,' *Journal of the American Medical Association*, 3 (1884), 41–2.

16 Gannett, 'Volume I,' p. 143 and p. 267, 'Volume II,' p. 15, 'Volume IV,' entry 11 May 1889 with letters tipped in.

17 J.W. Farlow, 'Obituary: William Whitworth Gannett, 1853–1929' *New England Journal of Medicine*, 200 (1929), 1177–80; C. Crenner, *Professional Measurement: Quantification of health and disease in American medical practice, 1880–1920* (Doctoral dissertation, Harvard University, 1993). Gannett, 'Volume I,' p. 315 with letter tipped in.

18 E.T. Morman, 'Clinical Pathology in America, 1865–1915: Philadelphia as a test case,' *Bulletin of the History of Medicine*, 58 (1984), 198–214.

19 Cabot, Patient Records, volume 18, p. 112 and volume 23, p. 8, for examples of Francis Lowell Burnett correspondence.

20 Copies of the 'Boston City Directory' and the 'Boston Telephone Directory' including the classified listings for the years 1910–1922, and 1913–1922 respectively can be consulted in the reference collections of the Boston Public Library.

21 It is useful to note too how concern over bureaucratic efficiency arose inside medical laboratories as well. Questions about the division of labor, quality control and work flow became defining features of the administration of large medical laboratories in the early-twentieth century. See P. Twohig, *Labour in the Laboratory: Medical laboratory workers in the Maritimes, 1900–1950* (Montreal: McGill-Queen's University Press, 2005).

22 I generalize cautiously, because I remember watching a physician in a private medical office in the 1990s duck into a back lab to add a little acid to a patient's urine sample for a quick idea about the degree of proteinuria, a practice that would have been immediately recognizable to Gannett.

5

Innovating Expertise: X-ray and Laboratory Workers in the Canadian Hospital, 1920–1950

Peter L. Twohig

Introduction

The modern healthcare system is a complex organization, comprising many occupational groups. There are those widely recognized as 'professionals' (physicians and nurses), those considered 'allied health professionals' (occupational therapy, physical therapy, speech pathology), and the 'support staff' (cleaners, porters, kitchen workers). Until recently, the historiography of healthcare in Canada has largely focused on professional groups and often uncritically. More recent studies, including the important work of Ruby Heap, Nadia Fahmy-Eid and others, have turned their attention to the allied healthcare groups.[1] While such studies have enriched our understanding of health services in Canada, the occupational group largely remains the focus of the analysis. This is entirely understandable. Many groups have membership records or publish journals that make them natural areas of inquiry. Nevertheless, one effect of this professional gaze is that occupational groups in healthcare have acquired what Gerald Larkin has described as an 'aura of inevitable permanence.'[2]

The history of technical work in hospitals presents an opportunity to disrupt notions of tightly contained occupational groups and explore the contested geography of healthcare work. Stephen Barley and Julian Orr recently described technical workers as 'the neglected workforce'[3] and, in the Canadian context, this is certainly true of laboratory and medical radiation technologists. Such workers are hidden from view as long as the technologies with which they work are operating. When problems occur, the technical workers are thrust into the limelight. As Jeffrey Keefe and Denise Potosky noted, technical workers became visible 'only when they made mistakes, departed from their assigned routines, or demonstrated incompetence.'[4] Technical workers ease our relationship with the healthcare technologies we encounter, including diagnostic imaging equipment. X-ray technologists, laboratory workers and others, such as sonographers, occupy the interstices between physicians and patients. Technical workers destabilize the history of discrete occupational groups and challenge traditional divisions between mental and manual work. The idea of clear, permanent and immutable professional boundaries becomes less clear through an analysis of hospital-based laboratory and X-ray workers.

The controversial case of Miss Gillis of Montreal

In the spring of 1928, a storm brewed in Sydney, Nova Scotia, Canada. Sydney was the chief city in industrial Cape Breton Island, an important coal mining area. Sydney and its environs underwent enormous population growth in the early decades of the twentieth century due to expansion in the coal and steel industry. In 1891, Sydney had a population of 2,427. Twenty years later the population was almost 18,000. Nearby Glace Bay, grew from a town of just less than 7,000 in 1901 when it was incorporated into the largest of all the coal towns a decade later. The population, which exceeded 16,500 in 1911, was also ethnically diverse, with perhaps about a dozen discrete neighborhoods in the town.[5] Economically, the 1920s was a difficult decade for Canada's Maritime Provinces (Nova Scotia, New Brunswick and Prince Edward Island) and economic hardship was particularly acute in Cape Breton. Coal miners there endured both wage cuts and layoffs, and generally the period was one of unprecedented strife between labor and capital. For example, the British Empire Steel Corporation attempted to reduce wages in 1922 by an astonishing 37.5 percent, prompting a strike. More conflict followed in the summer of 1923, early 1924 and, in March 1925, a five-month strike began, characterized by clashes between miners and company police and, ultimately, the use of the military. Declining economic conditions and the resulting unemployment led to the crisis in the coalfields.[6]

The gathering storm in the spring of 1928 did not concern another labor conflict between the working class and the coal or steel industry, but instead centered on the X-ray department at City Hospital.[7] The debate that unfolded in the press over several weeks between the hospital commission and the city council was about who had the ability to operate X-ray equipment and interpret radiographic findings. The hospital commission, supported by the medical staff and local physicians, requested $4,800 for new X-ray equipment. The motion also included a clause that would have given Dr C.M. Bayne, a radiologist, 50 percent of the earnings from the X-ray plant. Several city councilors and the mayor disapproved of such a scheme. Further proposals were put forward and the discussion at city council went long into the evening of 31 May. All the councilors agreed that new X-ray apparatus was required, in order to provide the patients with the latest treatment.[8] The city council understood the importance of X-ray equipment for the modern hospital. Indeed, the previous year the city approved $20,000 for the construction of a tuberculosis annex and had been asked to approve an additional $45,000 in 1928 to renovate facilities at City Hospital. Bayne argued that there was a profound need for the annex because 30 people died annually in the city each year from tuberculosis, while another 250 cases remained at large spreading the disease.[9] Surely, Bayne thought, there was a need for new diagnostic technology that would aid the tuberculosis control effort. Surely as well, his claim of half of the revenues was warranted, insofar as he would be lending his 'expertise' to the interpretation of X-ray plates, thereby helping his medical brethren in their struggle against TB.

Even though Bayne reduced his demand to 40 percent of revenues, Sydney Mayor James McConnell continued to speak against the plan. Indeed, the mayor believed that many of Sydney's doctors could read their own plates and would not need to consult with the radiologist at all. Moreover, the hospital had recently acquired a technician, Miss Gillis, from the staff of the Montreal General Hospital.[10] Gillis would be given 'full charge of the machine' and physicians would not have to operate the equipment, only interpret the results.[11] The *Sydney Post* reported that 'this young lady had been highly recommended ... for her efficiency and in any event her services were required for laboratory work.'[12] Gillis could 'take pictures' and provide plates directly to physicians, who would interpret them for their patients.

The hospital commission continued to urge for the appointment of Dr Bayne. They had the support of the medical staff of City Hospital who wanted a 'modern X-ray apparatus' and this included a physician to direct the use of the equipment.[13] Bayne had extensive experience, most recently at the provincial tuberculosis sanatorium in Kentville.[14] For their part, the hospital commissioners argued that across the Maritimes, sharing departmental revenues on a 50–50 basis was the norm.[15] The commissioners also noted that the 'operator' of the equipment (the physician) typically enjoyed the assistance of a 'technician.' The *Sydney Post* reported that the 'commissioners felt that although

Miss Gillis, Montreal X-Ray expert, comes highly recommended as a technician, she would not be qualified to interpret X-ray plates, or to give X-ray treatment.'[16]

The debate between the hospital commission and the city council intensified when the city council held a Saturday evening vote and struck the request for funding the X-ray equipment from the plebiscite ballot.[17] 'Acrimonious' debate divided the chamber for the next hour and, in the end, seven councilors voted to reject the proposal while six endorsed it. The *Sydney Post* reported that the issue was not whether the equipment should be purchased – everyone seemingly agreed on this point – but rather whether or not Bayne should be placed in charge. Alderman H.M. Israel typified the councilors contesting Bayne's appointment. Israel supported acquiring the new equipment but opposed placing it under Bayne's authority. Israel was quoted as saying 'While [council] do not know how to run a hospital, or an X-ray machine, we know how to run the City.'[18] For the opposition, placing the new X-ray apparatus under the direction of a medical man and sharing the revenues was not in the city's interest. Unstated, though implicit in his position, was that a trained operator had the requisite skills to utilize the new equipment. Less obvious, but equally important, was that the local physicians who would make use of the X-ray service *also* had the necessary skills to interpret the X-rays for their patients. A specialist such as Bayne was, in this view, an unnecessary extravagance.

Alderman Mack McLeod represented the other side of the debate. He gave a 'forceful address' to city council, noting that the X-ray equipment at City Hospital was woefully inadequate and recently condemned by a Halifax expert. McLeod described the equipment evocatively as a 'miniature lightening storm' adding that 'many citizens have gone and spent time and money waiting to get photographs, when it would have paid them better to go to a local photographer.'[19] McLeod noted that those in favor wanted new equipment, a good technician to operate the equipment and an expert to interpret the radiographic findings. For McLeod, all three elements were necessary for the hospital to be modernized. Nor was the debate really about Miss Gillis. McLeod commented in passing that she had been appointed against the wishes of the medical staff, which touched off another row. Mayor McConnell rejected this outright, saying that no physician had opposed the appointment of Gillis, 'So help me God.'[20] Indeed, Gillis' credentials or qualifications were never called into question and there was general agreement among the councilors and in the press that she would make a very fine addition to the X-ray department. Rather, the debate was about what was required to bring the department up-to-date and thereby modernize the hospital. The acquisition of new equipment was only a useful first step. What was really necessary to complete the modernization project was the addition of a technician to operate the equipment and an expert to interpret the results for local doctors.

The debate continued to unfold in the press over the next several weeks. An editorial that appeared on 5 June 1928 described the opposition to Bayne's appointment as little more than a display of 'pettishness and autocracy.' The *Post* continued:

> That the electors of Sydney should be denied the right of pronouncing judgment on so vitally important a question as the installation of an up-to-date, not to say a safe, X-ray system in the City Hospital, simply because Mayor McConnell and several of the Aldermen want a graduate nurse instead of a practicing medical specialist such as Dr Bayne, to operate the apparatus and direct the technical work of the department, discloses a state of affairs which should arouse the people of this city ...[21]

Subsequent articles noted that a 'well-trained technician can only take films and develop them and assist in preparing the patient for X-ray.' The technician could not interpret the plates. Nor could technical hands perform fluoroscopes or use the X-ray apparatus for therapy. The result, according to the *Sydney Post*, would be underutilized equipment. 'If the City of Sydney is going to expend [money] to modernize the now obsolete X-ray plant at the City Hospital,' the paper argued 'then in justice to the public, the machine must do all that it is capable of, its uses not restricted to the taking of plates alone, but intended to embrace all types of X-ray work ... The machine itself cannot do this, a graduate nurse who is a thoroughly trained technician cannot dot [sic] it, but only a medical man who has a thorough knowledge and experience.'[22] Ultimately, the hospital commission rescinded the Bayne appointment and the City Council unanimously approved of the acquisition of new X-ray equipment for Sydney's City Hospital.[23] Gillis took charge of not only the X-ray department but also the laboratory at the hospital.[24]

The excitement generated by Miss Gillis' appointment, which centered on whether she had the requisite skills to operate the X-ray equipment at the hospital, provides a useful illustration of how uncertain roles were within the modernizing healthcare system. Gillis was a

nurse but was appointed to City Hospital to staff the X-ray and laboratory departments. Her appointment also touched off a debate regarding
whether X-rays could be interpreted by all doctors or whether they
required a consultation with a specialist, such as Dr Bayne. Such
debates illustrate the contested geography of healthcare work.[25] That is,
the scope of practice of occupational groups within healthcare was still
very much a matter to be settled. Many hospital workers in Canada
filled multiple roles in the modernizing hospital. X-ray workers, laboratory personnel, kitchen staff and clerks were rarely confined to one
aspect of hospital labor, while nurses often found themselves working
throughout the hospital across many different services.[26]

X-ray workers

Debates about expertise and skill were common in hospital work in
Canada's Maritime provinces, and elsewhere, as the broader social
history of the creation of X-ray departments reveals. Within months of
their discovery in November 1895, X-rays were recognized as a powerful imaging tool and were being used for clinical purposes in Canada
and the United States.[27] In his classic study of medical technology,
Stanley Reiser wrote that 'No previous diagnostic discovery had stirred
quite so much public interest and involvement as the X-ray.'[28]
Physicians and hospitals recognized that X-rays and radiation therapy
were clinically useful and full of therapeutic promise. Roy Porter's
description is particularly apt: 'What the stethoscope had been to the
nineteenth century, the X-ray became for the twentieth: an impressive
diagnostic tool and a symbol of medical power.'[29] In his history of the
Toronto General Hospital, J.T.H. Connor argues that the acquisition of
X-ray technology (used first in November 1896) 'serves as the best
example of medicine reflecting modernity.'[30]

X-rays are, then, symbolically important as one of the markers of the
modern hospital. Certainly, many historians of healthcare point to the
adoption of such technologies as evidence of this modernization. But a
closer examination reveals that a good deal still needed to be sorted
out, particularly with respect to staff. In other words, ideas of expertise
still needed to be innovated. And, despite the emphases of Porter and
Connor on the meaning of X-rays for medicine, many different people
operated X-ray equipment. In his formidable study of American hospitals, Charles Rosenberg has described how 'photographers, clinicians,
and technicians' ran X-ray departments in various settings.[31] Other settings show a similar diversity. In England, a dentist operated the X-ray

apparatus at the Nottingham General Hospital, while at London's St Mary's Hospital, the 'theater beadle' imaged fractured limbs.[32] Similar patterns may be discerned in Canada's Maritime Provinces. When the St John Infirmary opened in Saint John, New Brunswick in 1914 it was a local general practitioner who supervised the X-ray work.[33] In Nova Scotia, early X-ray work 'was relegated to the physicist and photographers – medical men would have nothing to do with what was termed "machine made diagnosis",' according to prominent Nova Scotian radiologist Herbert R. Corbett.[34]

Given this variability, it is hardly surprising that in many settings, day-to-day operations of the X-ray departments fell to nurses. Recruiting staff to X-ray departments, either as assistants or to take charge of the service, was an important question as the departments were transformed during the 1920s from small services to important departments within hospitals, with huge volumes of work. When Dr William H. Eagar was appointed roentgenologist in Halifax's Victoria General Hospital in November 1919, he was promised that the hospital would supply 'an assistant, not necessarily a so-called technician, who shall be an employee of the nursing department.'[35] The nurse, when working in the X-ray plant, was under Eagar's direction, who assumed responsibility 'to train and develop' her for the work.[36] When an early candidate declined the position,[37] Eagar plucked a senior male nurse from the staff to train as a 'technician' in late January 1920.[38]

Many nurses recalled working in X-ray departments in oral histories collected during the 1980s by nursing historian Barbara Keddy.[39] Greta MacPherson, who began her career at Glace Bay General Hospital in the summer of 1922, worked in both the X-ray department and the laboratory following her graduation, replacing another nurse. Both of the nurses took a 'short course' in X-ray and lab work in Halifax. Young Greta, who gave up school teaching to become a nurse, was not enthusiastic about working in the diagnostic service departments, and worried that her duties there would interfere with her desire to nurse. She was, however, adept in X-ray and laboratory technique and the hospital superintendent assured her that she would continue to nurse patients. MacPherson recalled that the superintendent 'was anxious for me to do this. Nobody would touch it. So I took instruction.' Over the next ten years, her duties at Glace Bay General combined nursing with X-ray work, basic laboratory tests such as blood counts and urine samples, and administering anesthetics. She came to particularly enjoy her work in the X-ray department. Probably echoing the opinion of small hospital administrators everywhere, MacPherson stated 'to be a nurse-technician, you're a jump ahead of when you've just taken a technician's course because you already know how to handle patients.'[40] Her ability to work across several departments and fill multiple roles permitted the hospital to undergo a degree of modernization, without the burden of adding dedicated staff members to now-required departments such as laboratory or X-ray services. Many other nurses recounted similar stories.[41]

Flora McDonald, who trained in Glace Bay between September 1928 and September 1930, recalled that student nurses staffed the general hospital almost exclusively. There was a superintendent and an assistant and, she noted 'they had a lab technician and an X-ray technician – a lab and X-ray, and when we went in first they did them both.' MacDonald vividly captured the multi-faceted duties that nurses were required to fulfill:

> Well, each Sunday ... [the nurses] saw our sick patients first ... And if it was a busy day – now I remember one Sunday, now I'm not saying I'm the only one who could do this, anyone could do it, we knew how to run the portable X-ray, so when a minor accident came in or some emergency we went down and admitted the patient, and we could do urine for sugar and albumin, we could do a microscopic, and we had to run the portable X-ray ... You were everywhere. And you could handle anything. And that's not just one staff member – that's what we were taught to do. ... So for that size hospital, you get an excellent training, and I remember [the superintendent] telling us when we went away to do post-graduate, 'Don't ever be afraid of going from a smaller hospital to a larger one'.[42]

Another nurse who took up work in the X-ray department was Clara McKinnon. Her mother enrolled McKinnon in nursing school, while young Clara was living in Michigan with an older married sister. She was told to return home and began her nurses' training at Glace Bay General Hospital on 12 October 1929. After graduation, McKinnon worked in the X-ray department. The nurse who was in charge took three months off to study for an examination, so McKinnon was placed in charge of the X-ray equipment. She had spent some time in the department while training. She remembered that the hospital superintendent, Miss McMillan, believed that nurses:

> should be able to go do everything, every department of the hospital and fill in. Whether it's the kitchen, whether it is ... even the

furnace room. You should know where the furnace room is and how to turn it off or whatever is wrong and you should know the lab, the X-ray, the pharmacy, the laundry, the whole business. Make yourself aware that you know that you can fill in anywhere if you're in the hospital. ... Jack-of-all-trades.[43]

Mary Murphy recalled a similar experience at the New Waterford General Hospital, where she began as a student in September 1925. Murphy, the daughter of a steelworker and a dressmaker, recalled that nurses were responsible for taking X-rays. She particularly remembered that Miss A.J. MacDonald, who worked in the laboratory and took X-rays, 'was really something. It was funny, she took to me and she used to give me a lot of training in the lab. And I'll never forget what she taught me, she taught me to take X-rays.' Murphy would spend 20 years 'taking pictures', suggesting that for some nurses, working in settings such as X-ray departments was neither peripatetic nor temporary.[44]

Laboratory workers

Laboratories were established as part of the infrastructure of healthcare in the late-nineteenth and early-twentieth century. Like X-ray departments, they are typically viewed as a pillar of healthcare's 'modernization,' which also encompassed new legislation, a dramatic period of hospital construction, the transformation of Canadian medical schools, and the creation of nursing programs in hospitals and in universities. The growth of laboratories was greatly facilitated by the federal government's efforts to control what has become widely known as the 'venereal disease problem.' Most provinces passed legislation that addressed VD in the years before 1920 and venereal disease control became an integral part of the federal health department when it was established in 1919. Two hundred thousand dollars was devoted to controlling the 'secret plague.'[45] The federal government would supply the funds in proportion to provincial population, while the province would deliver the services.

The impact was dramatic. In Nova Scotia's public health laboratory, 610 tests were conducted during 1919–20, which grew to more than 2,500 only 2 years later. From 1920 to the mid-1930s, the number of syphilis tests conducted in the laboratory increased 16-fold.[46] The rise of venereal disease testing is only one clinical example of how laboratory work grew in the 1920s. There were others. Indeed, at the

Pathological Institute in Halifax, which did both clinical laboratory tests for the Victoria General Hospital and public health tests for the province, the number of tests increased from 759 in 1914–15 to close to 9,000 a decade later and exceeded 12,000 tests by the end of the 1920s.[47] Smaller community hospitals also began to establish laboratories in order to gain accreditation. Conditions were ripe for a dramatic expansion of the laboratory staff.

While the demand for workers capable of conducting laboratory tests was strong, there were no formal training programs. Many hospitals relied upon existing staff and, often, sent nurses to learn basic laboratory techniques. The courses were short, ranging from two- to eight-weeks in the 1920s.[48] While their training was modest, these nurses often worked with a high degree of independence. In early 1931, when Prince Edward Island's (PEI) health department was searching for someone to fill a position in their new laboratory, they considered a number of candidates, including two female physicians.[49] Ultimately, the health department settled on Marion Merry, who had five years' experience at the Toronto General Hospital, where she was responsible for bacteriological and pathological tests. PEI's health officer, Dr P.A. Creelman acknowledged with remarkable frankness that a laboratory technician would have to work very independently because 'I do not think there will be anyone here qualified to give her much supervision.'[50] In February 1931, R.E. Wodehouse, the executive secretary of the Canadian Tuberculosis Association, expressed the same view, writing to the prospective candidate 'you will mostly be your own boss.'[51] Clearly, Wodehouse and Creelman thought that a woman with five years' laboratory experience was fully capable of assuming responsibility for the laboratory's day-to-day operations.

When Merry ultimately rejected the appointment, Wodehouse contacted Dr Robert Defries of Connaught Laboratories to see whether there was another suitable candidate.[52] After further consultation with Dr John FitzGerald, another of Toronto's leading public health figures, Wodehouse decided on another path. Rather than try to recruit a trained technician from Toronto, or another center, a Prince Edward Island nurse would be sent to Halifax to train at the Pathological Institute. Wodehouse wrote to Benjamin C. Keeping, the assistant medical health officer for Prince Edward Island that:

> I discussed the matter in Ottawa Wednesday with Dr Fitzgerald at considerable length and he thought we would be well advised to take this course, because, if the nurse has friends in Charlottetown

she would be socially at home there. He felt that she likely would be more suitable than someone strange to the Island whom we might find to take there, and he also felt that in the long run that a contented person with long service would be of greater advantage to us than an over-efficient person to begin with who might become restless and leave us at the end of nine or twelve months.[53]

The PEI health department ultimately found another woman, Esther Stevenson, to fill the job. She was a graduate of Prince of Wales College, who taught school on Prince Edward Island for four years. In 1922, she began training as a nurse in Massachusetts. She undertook further training, including some brief training in basic laboratory work, where she learned how to do urinalyses and stain sputum samples for the tubercle bacillus. She also attended lectures in bacteriology, one of the fundamental sciences of public health work, but did not have experience in this area.[54] While her laboratory training was modest, she was an Island nurse willing to undertake further laboratory training. Staffing decisions were shaped by many variables, including those of home or place in the above example, religion or local employment conditions.[55] Skill and efficiency were not the only considerations when filling positions in the emerging healthcare complex.

Esther Stevenson, despite undertaking training in Halifax, never assumed the position in the laboratory, opting instead to be married.[56] Nevertheless, her selection is instructive for a number of reasons. First, there was the important role of the Canadian Tuberculosis Association in providing funding, part of a broader pattern that prevailed in the Maritimes. Innovations in healthcare from the 1920s through to the 1960s were often funded through federal-provincial initiatives, philanthropic and volunteer activities, Maritime-wide strategies, provincial-municipal cooperation, or coordinated services across municipal boundaries. Indeed, this is one of the most intriguing features of health services in the first half of the twentieth century. There was also the important question of attempting to hire a person with local connections. Stevenson's appointment reveals that staffing decisions were not made solely on the basis of formal credentials. PEI's public health officials recognized that she was not the most qualified person. Equally enlightening is the variability of the women considered for the position. There were two women who had laboratory training, a number of nurses, and two female physicians considered for the public health laboratory. Moreover, relief and assistance in the laboratory was to be provided by the public health nurses.[57] All were thought to have the required skills to take charge of the lab, even in the absence of a supervisor. After all, the successful candidate, according to the physicians, was to be 'her own boss.' Thus, while Margarete Sandelowski describes the nurses working in diagnostic services as 'assistants'[58] they were often working independently in many settings.

Fluid roles in hospitals

The history of technical workers in Canada offers important insights into the nature of hospital work. X-ray and laboratory workers reveal a pattern of working across services, wherein their occupational roles were highly fluid. Further evidence of their multiple roles may be gleaned from advertisements for jobs, or from individuals seeking work, in Canadian healthcare periodicals. Pearl Morrison wrote in *Canadian Nurse* in 1941 that graduate nurses often worked in the laboratory or X-ray departments.[59] Anne Wright, writing a year later, made the same point but also noted that 'X-ray equipment has become so greatly simplified and reduced in cost that there are few of the smaller hospitals now with it, but its operation remains a problem. If the hospital can afford a technician and a part-time radiologist, or a part-time technician who may divide his or her time with other departments, the difficulty is overcome.'[60]

Many settings provided training intended to meet the demand for a multitasking labor force. Dr J.C. McMillan, director of the Winnipeg General Hospital's radiology department, advertised a 16-month course approved by the American Registry of X-ray Technicians that was only open to nurses.[61] The reality was that many hospitals foisted multiple duties upon existing staff, as illustrated in this 1942 article in *Canadian Hospital*:

> Since in X-ray work so much depends upon the proper handling of sick people, it is reasonable to suppose that nurses would make good technicians, and it often works out that way. It is obvious that they have already had much of the professional attitude and responsibilities taught to them. ... It has become a common practice – which is working very well in many small hospitals – to have one person do both laboratory and X-ray work.[62]

Advertisements for jobs, or from individuals seeking work, provide further evidence. An unnamed eastern Canadian hospital advertised for a 'Laboratory X-ray technician, able to do blood chemistry' in

1930.[63] The next year, a woman advertised that she spent eight years working as a 'nurse-Laboratorian' in a doctor's office, 18 months as a combined X-ray and laboratory technician in a 75-bed hospital and another year working exclusively in the X-ray department of a 350-bed hospital.[64] Early issues of the *Focal Spot*, the official journal of the Canadian Society of Radiological Technicians that began publishing in 1944, offer abundant evidence. A Manitoba woman seeking a position in a large urban hospital emphasized that she was a registered nurse and a member of both the Canadian and American X-ray societies. A nurse from Alberta, who worked first for five years performing lab tests, was now currently working as a combined X-ray and laboratory technician.[65] The St Catharines General Hospital, in southern Ontario, sought an X-ray technician, though preference would be given to one with nurses' training. The Woodstock General Hospital, also in Ontario, preferred an individual who could work both in the laboratory and in the X-ray room.[66] One hospital wanted a nurse to work in the X-ray and laboratory departments, noting that the candidate would spend their 'spare time' as a general duty nurse, while another advertised for a nurse technician who would maintain medical records in her spare time.[67] Focusing exclusively on a single aspect of an individual's labor, as many studies of healthcare professionals are wont to do, obscures the complex nature of hospital work, wherein many worked across disciplines and many employers demanded broad and flexible skills. Moreover, this was an enduring feature of hospital work. In Halifax, for example, the Tuberculosis Hospital employed a combined X-ray and laboratory technician as late as 1949.[68] Laboring across services was standard fare for X-ray workers and others in the inter-war period and beyond, in rural and urban settings, in large and small facilities.[69]

It is clear that women were critical to the expansion of diagnostic services in Canadian hospitals. The analysis of technical workers in the Maritimes, and the national patterns that these illustrate, also reveals the multiple roles of women in hospitals. Women's labor was, to put it bluntly, a significant factor in the making of the modern hospital. In her analysis of auxiliary workers in dentistry, Tracey Adams has convincingly illustrated such workers aided the efforts of dentists to professionalize, in part because they permitted dentists to distance themselves from 'lower-status aspects of their work.'[70] Like dental assistants, laboratory workers and X-ray technicians were 'channeled into support occupations' that, ultimately, supported the expansion of diagnostic services within the hospital.

At the same time, however, their position as interchangeable women workers, capable of filling in when necessary, limited their claims for full status as healthcare professionals and ensured their subordinate status to medicine. Women performed a wide variety of tasks in the Canadian hospital and played an integral role in its scientific and technological transformation and, through it, of the healthcare system. That many of the earliest laboratory and X-ray workers were women is hardly surprising. While nurses used the new technologies in the diagnosis, management and, in the case of X-rays, treatment of patients, they did so in the service of both physicians and hospital administrators, who were eager to bring the latest developments to their patient population. As Sandelowski recently wrote, doctors viewed nurses 'much like stethoscopes and surgical instruments, as physical or bodily extensions of physicians.'[71] 'In order to harness the benefits of this new technology,' Sandelowski argued, 'physicians had to share its use with nurse (and eventually also with a host of new technicians whose jobs were created in response to it).'[72] In the new scientific hospital of the twentieth century, patients routinely experienced blood tests or urine samples and physicians now depended upon others to collect the samples, take the images, or prepare reports.

What these technical workers could not do was interpret the results or make diagnoses. Diagnosis, the mental component of clinical care, was in this way separated from the manual aspects that underpinned that diagnosis and vested in the professional domain of physicians. Though many hospital workers were expected to assume responsibility for running the new services, often with little or no supervision, they did not derive much in the way of cultural or professional authority from the technology. Moreover, as Rosemary Stevens has argued, the multifaceted nature of nursing work further defined nursing as 'all-purpose female service workers without a defined monopoly of scientific skills.'[73] Her argument fits the experience of other multitasking women in healthcare. Physicians used stethoscopes to diagnose disease; nurses used them to collect information which was then passed to the physician for interpretation. Women's engagement with technology in the diagnostic services, rather than providing a fresh impetus to professional claims, served instead to blur their role, while concurrently confirming their subordinate position to physicians.

It is not entirely clear how nurses themselves felt about their use of technology and their new duties within the hospital. Lavinia Dock, a leading American nurse, recognized that opportunities in the service departments could alleviate some of the overcrowding that was characteristic in American nursing as early as the 1890s. Dock suggested that departments such as dietetics or pharmacy were promising employment alternatives for nurses. Moreover, Dock believed such services would be better served through staffing them with nurses.[74] Rank and file nurses expressed ambivalence toward these new roles. Some nurses, such as Greta MacPherson, feared new duties would interfere with their nursing work. Greta, after all, suggested that nobody else wanted to do the work, evidence that such positions were not desired.

The technical workers who assumed their place in the healthcare system suggests that we need to rethink some of our assumptions about specialization in healthcare and how workers were prepared for practice. There is, furthermore, a profound need to situate interpretations of professional and occupational groups within the historical and structural parameters of the development of health services and, through this, grounding them within the larger processes of welfare capitalism. The experience of X-ray and laboratory staff reveals a portrait of worker's multiple duties. Such patterns of work are revealed through detailed analysis of individual workers in particular settings. Rather than threadbare stories of professional formation, an approach that focuses on the complex and contradictory roles of healthcare workers yields new insights that more traditional understandings of single occupational groups obscure or entirely omit. Most intriguing are the ways in which healthcare workers continued to work across services long after the presumed specialization of Canadian hospitals. In other words, the history of the hospital looks different when viewed from the perspective of technical workers, multitasking nurses or myriad other employees who labored in the new departments or services.

Notes

1 R. Heap, 'Physiotherapy's Quest for Professional Status in Ontario, 1950–80,' *Canadian Bulletin of Medical History*, 12:1 (1995), 69–99 and Heap, 'Training Women for a New "Women's Profession": Physiotherapy Education at the University of Toronto, 1917–40,' *History of Education Quarterly*, 35:2 (1995), 135–58. In Quebec, there have been studies of a number of allied healthcare workers. See, for example, L. Piché and N. Fahmy-Eid, 'À La Recherche d'un Staut Professionnel dans le Champ Paramédical: Le cas de la diététique, de la physiothérapie et de la technologie médicale,' *Revue d'Histoire de l'Amerique Francaise*, 45:3 (1992), 375–401; N. Fahmy-Eid and L. Piché, 'Le Savoir Négocié: Les stratégies des associations de technologie médical, de physiothérapie et de diététique pour l'accès à une meilleure formation professionnelle (1930–1970),' *Revue d'Histoire de l'Amerique Francaise*, 43:4 (1990), 509–34; A. Charles and

N. Fahmy-Eid, 'La Diététique et la Physiothérapie Face au Problème des Frontières Interprofessionnelles (1950–1980),' *Revue d'Histoire de l'Amérique Française*, 47:3 (1994), 377–408. For a broader discussion of healthcare workers in Québec, see N. Fahmy-Eid, *Femmes, Santé et Professions: Histoire des diététistes et des physiothérapeutes au Québec et en Ontario, 1930–1980* (Saint-Laurent: Fides, 1997).

2 G. Larkin, *Occupational Monopoly and Modern Medicine* (London: Tavistock Publications, 1983), vi.

3 S.R. Barley, and J.E. Orr, 'The Neglected Workforce,' in Barley and Orr (eds), *Between Craft and Science: Technical work in US settings* (Ithaca: Cornell University Press, 1997), 1–19.

4 J. Keefe and D. Potosky, 'Technical Dissonance: Conflicting portraits of technicians,' in Barley and Orr (eds), *Between Craft and Science*, 54–5.

5 For information about public health in this part of Nova Scotia, see P.L. Twohig, 'Public Health in Industrial Cape Breton 1900–1930s,' *Royal Nova Scotia Historical Society Journal*, 4 (2001), 108–31.

6 The history of labor conflict in industrial Cape Breton during the 1920s has been well told. D. Macgillivray, 'Military Aid to the Civil Power: The Cape Breton experience in the 1920s,' *Acadiensis*, 3 (1974), 45–64; D. Frank, 'The Cape Breton Coal Industry and the Rise and Fall of the British Empire Steel Corporation,' *Acadiensis*, 7 (1977), 3–34; idem, 'Class Conflict in the Coal Industry: Cape Breton 1922,' in G.S. Kealey and P. Warrian (eds), *Essays in Canadian Working Class History* (Toronto: McClelland and Stewart, 1976), 161–84; idem, *J.B. McLachlan: A Biography* (Toronto: James Lorimer & Company Ltd., 1999). For broader studies of the Maritimes that examine the economy, see J.G. Reid, *Six Crucial Decades: Times of change in the history of the Maritimes* (Halifax: Nimbus, 1987); I. McKay, 'The 1910s: The stillborn triumph of Progressive Reform' and D. Frank, 'The 1920s: Class and region, resistance and accommodation,' both in E.R. Forbes and D.A. Muise (eds), *The Atlantic Provinces in Confederation* (Toronto: University of Toronto Press, 1993), 192–229 and 233–71.

7 'City Hospital Opened,' *Sydney Post* (16 November 1915).

8 'Appointment of Doctor Deferred,' *Sydney Post* (1 June 1928). The 50-percent interest was not unusual in Nova Scotia. In 1920, Victoria General Hospital Superintendent W.W. Kenney sent letters across Canada inquiring about the method of paying physicians in charge of X-ray departments. See, for example, W.W. Kenney to Dr James C. Fyshe, Royal Alexander Hospital, Edmonton, 14 June 1920, in Victoria General Hospital Letterbook (hereafter VGHL). This important collection of hospital records dating from the mid-nineteenth century was once held at the provincial archives (RG25, Series B, Section 1) but subsequently removed from the public archives and returned to the Queen Elizabeth II Health Sciences Centre (QEII-HSC), in Halifax.

9 Sydney City Council Minutes (hereafter Sydney Minutes), 8 May 1928, Beaton Institute, University College of Cape Breton, Sydney, Nova Scotia.

10 I could not identify this woman precisely. There are several possibilities listed in the 1928 *City Directory* for Sydney, North Sydney, Sydney Mines, Glace Bay, Dominion, New Waterford and Louisbourg. In Glace Bay, there was a nurse named Annie Gillis, who lived at 1132 Main Street in Glace

Bay. It is unlikely, however, that she was the Miss Gillis working at City Hospital. In Sydney, Veronica and Catherine Gillis lived together at 233 Rockdale Street, both of whom worked at the hospital. There were no other Miss Gillis' listed in the other communities who worked as nurses or at the hospital. But there is always the possibility that the X-ray department's Miss Gillis was simply not listed in the directory.

11   'Appointment of Doctor Deferred,' and Sydney Minutes, 31 May 1928.

12   'Appointment of Doctor Deferred.'

13   'Need for New X-ray Equipment is Stressed,' *Sydney Post* (2 June 1928).

14   In 1900, Nova Scotia's legislature passed legislation to establish the provincial sanatorium and, when it opened in 1904, it was the first state-operated facility in North America. The best study remains S.M. Penney, *Inventing the Cure: Tuberculosis in 20th century Nova Scotia* (PhD dissertation, Dalhousie University, 1991). For a popular history, see D.F. Ripley, *Thine Own Keeper: Life at the Nova Scotia Sanatorium, 1904–1977* (Hantsport: Lancelot Press, 1992).

15   'Need for New X-ray Equipment Is Stressed,' *Sydney Post* (2 June 1928).

16   *Ibid.* The article explicitly noted that using X-rays for treatment was 'a very important part of measures in combating disease, and as the law does not allow the indiscriminate use of this sensitive treatment by other than medical men, the commissioners felt that such a valuable remedy should be available for the Cape Breton public.' In other words, a physician needed to be in charge of the equipment.

17   Incorporated towns and cities in Nova Scotia were required to hold plebiscites if they wished to borrow money. The plebiscite ballots were approved by municipal governments and voted upon by ratepayers. In earlier plebiscites in Sydney, for example, ratepayers voted to approve expenditures to make improvements to the sewage system and to finance the construction of a hospital. Sydney Minutes, 3 July 1902, 21 July 1903, 26 January 1913.

18   'X-Ray Vote Is Refused By City Council,' *Sydney Post* (4 June 1928).

19   *Ibid.* The 'Halifax expert' was Dr S.R. Johnston, considered one of the leading X-ray authorities in the province.

20   *Ibid.*

21   'Autocracy At City Hall,' *Sydney Post* (5 June 1928).

22   'The Importance of X-Ray and Its Uses,' *Sydney Post* (6 June 1928). Subsequent editorials (6 June and 7 June) took square aim at Mayor McConnell's position, describing the position of the councilors opposed to Bayne's appointment as 'the rawest kind of parochialism, unworthy of the business administrators of a modern community.'

23   'City Hospital To Have A New X-Ray,' *Sydney Post* (21 July 1928).

24   Beatrice Andrews, 'Annual Report of City of Sydney Hospital,' in City of Sydney Annual Reports 1928, Beaton Institute.

25   I would like to acknowledge the essential critical advice of Dr Robert C.H. Sweeny, Memorial University of Newfoundland, who has offered many helpful comments on aspects of this work and the idea of contested geographies in particular.

26   I have explored this issue through a case study of Maritime laboratory workers. See P.L. Twohig, '"Local Girls" and "Lab Boys": Gender, Skill and

Medical Laboratories in Nova Scotia in the 1920s and 1930s,' *Acadiensis*, 31 (2001), 55–75 and P.L. Twohig, *Labour in the Laboratory: Medical laboratory workers in the Maritimes* (Montreal and Kingston: McGill-Queen's University Press, 2005), especially chapters 4 and 5.

27 The centenary of Röntgen's discovery prompted a spate of writing in journals such as *Journal of the American Medical Association, British Medical Journal, Canadian Medical Association Journal* and others. There are also several studies that focus on Canada. The fullest treatment is J.E. Aldrich and B.C. Lentle (eds), *A New Kind of Ray: The radiological sciences in Canada 1895–1995* (Vancouver: Canadian Association of Radiologists, 1995). Other work includes R. Brecher and E. Brecher, *The Rays: A history of radiology in the United States and Canada* (Baltimore: Williams and Wilkins, 1969) and E.A. Shorter, *A Century of Radiology in Toronto* (Toronto: Wald and Emerson, 1996). C.R.R. Hayter has been the most prolific scholar, providing many detailed studies of aspects of X-ray work. For a sampling, see C.R.R. Hayter, *An Element of Hope: Radium and the response to cancer in Canada, 1900–1940* (Montreal and Kingston: McGill-Queen's University Press, 2005); idem, 'The Clinic as Laboratory: The case of radiation therapy, 1896–1920,' *Bulletin of the History of Medicine*, 72 (1998), 663–88; idem, 'Making Sense of Shadows: Dr. James Third and the introduction of X-rays, 1896 to 1902,' *Canadian Medical Association Journal*, 153 (1995), 1249–56. Hayter has also examined Nova Scotia in '"To the Relief of Malignant Diseases of the Poor": The acquisition of radium for Halifax, 1916–1926,' *Journal of the Royal Nova Scotia Historical Society*, 1 (1998), 130–43.

28 S.J. Reiser, *Medicine and the Reign of Technology* (Cambridge: Cambridge University Press, 1978), p. 62.

29 R. Porter, *The Greatest Benefit to Mankind: A Medical History of Humanity* (New York: Norton, 1997), p. 606. It is of course worth noting that the interest in and authority of X-rays quickly spawned a host of dubious X-ray 'cures' and the widespread use of X-rays for such diverse purposes as the treatment of minor benign menstrual bleeding or appropriate sizing of feet. See *ibid.*, p. 608 for the first example and J. Duffin and C. Hayter, 'Baring the Sole: The rise and fall of the shoe-fitting fluroscope,' *Isis*, 91 (2000), 260–82 for the use of X-rays in shoe stores. Importantly, the use of X-rays was done with little regard to patients. One consequence of the therapeutic use of X-rays to treat benign menstrual bleeding, for example, was cervical cancer. Technical staff in particular was put in harm's way, often with devastating consequences. Reiser also describes the widespread use of X-rays for 'sentimental' photographs, showing lovers clasped hands or fashionable women's bejeweled fingers. Reiser, *Medicine and the Reign of Technology*, 60–3.

30 J.T.H. Connor, *Doing Good: The Life of Toronto's General Hospital* (Toronto: University of Toronto Press, 2000), 130. References to the early use of X-ray equipment are included in a chapter tellingly entitled 'A Model Hospital.'

31 C.E. Rosenberg, *The Care of Strangers: The rise of America's hospital system* (Baltimore: Johns Hopkins University Press, 1995), 182–3. It is also worth noting that there were also competitors outside of the hospital. See C.C.S. Murphy, 'The Control of Medical X-Rays, 1895–1917: Why scientists as paramedics, not medics as parascientists,' *Social History of Medicine*, 40 (1987), 67–8.

32 R. Stevens, *Medical Practice in Modern England: The impact of specialization and state medicine* (New Haven: Yale University Press, 1966).

33 *75 Years of Caring: St Joseph's Hospital* (Saint John, New Brunswick), n.p. [1989].

34 Herbert R. Corbett, 'Inter-Relationships,' *Focal Spot*, 3 (1946), 132–3.

35 Minutes of the Victoria General Hospital Board of Commissioners (hereafter BOC), 16 May 1914 and Minutes of the Medical Board (hereafter MMB), 24 November 1919. Eager stayed in this position until 1926, see BOC, 13 August 1926.

36 W.W. Kenney to W.H. Eagar, 10 November 1919 in VGHL. In correspondence with the Hon. E.H. Armstrong, the Minister of Public Works and Mines, Kenney wrote that the VG 'will supply a *technician* [my emphasis] to this department ...' Kenney to Armstrong, 26 November 1919 in VGHL.

37 W.W. Kenney to Mary Noonan, 20 January 1920, VGHL.

38 W.W. Kenney to Dr A.F. Miller, Superintendent, Nova Scotia Sanatorium, 27 January 1920 in VGHL. While unnamed, it is likely that the man was Michael MacInnis, who would serve as X-ray technician well into the 1940s. See BOC, 2 June 1943.

39 This material is drawn from the Barbara Keddy Fonds, Series 018, Social History of Nursing in Nova Scotia in the 1930s, Nova Scotia Archives and Records Management (NSARM). I am grateful to Dr Keddy for her permission to cite from this important collection.

40 Keddy Fonds, MF160-11, Interview with Greta MacPherson. When Keddy asked MacPherson about radiation exposure from doing the X-ray work, MacPherson replied that 'I always blamed that for my infertility,' suggesting the hazards associated with this work.

41 While nurses were obviously important to the operation of many X-ray departments, others continued to operate the equipment. In the early 1920s, Wendall Bain served as the maintenance man at the Yarmouth Hospital, before going to Halifax to take an X-ray course. He returned to the hospital and served as the head of the X-ray department for two and a half years. E.R. Pothier, *Mary Ann Watson and the Yarmouth Hospital* ([Yarmouth]: s.n., [1986]).

42 Keddy Fonds, MF 160-10, Interview with Flora K. McDonald.

43 Keddy Fonds, MF 160-4, Interview with Clara M. Buffet.

44 Keddy Fonds, MF 160-27, Interview with Mary Kathleen Murphy.

45 J. Cassel, *The Secret Plague: Venereal disease in Canada, 1838–1939* (Toronto: University of Toronto Press, 1987), 163–9.

46 Nova Scotia Public Health Annual Report (hereafter PHAR) 1919–20 to 1929–30. The laboratory reported doing some 159 Kahn tests in 1925–6, and this number grew to over 2,700 the next year, outpacing the 2,369 Wassermann's completed. Thereafter, only Kahn's were conducted, topping 4,000 in 1928–9, almost 7,000 in 1930–1, 9,000 in 1933–4 and reaching 10,000 in 1934–5.

47 These numbers are based upon data published in the *Annual Reports* and reflect the number of tests, not the number of samples. The totals are my own. Despite the imperfections of these data, they do serve to illustrate the significant growth in the work of the laboratory.

48 This is developed more fully in Twohig, '"Local Girls" and "Lab Boys"'.

49 National Archives of Canada, Canadian Tuberculosis Association (CTA), MG 28, I 75, File 54. Creelman to Wodehouse, 29 December 1930.
50 CTA, File 54. P.A. Creelman to Dr D.J. MacKenzie, n.d. [March 1931]. This letter is a copy of the original.
51 CTA, File 54, Wodehouse to Marion Merry, 12 February 1931.
52 CTA, File 54, Wodehouse to Defries, 24 February 1931.
53 CTA, File 54, Wodehouse to Keeping, 27 February 1931.
54 CTA, File 54, Stevenson to Creelman, 23 January 1931. Stevenson had previously expressed some interest in securing a position on Prince Edward Island. She had applied for, and apparently was promised a nursing position at the Provincial Sanatorium. See CTA, File 54, Creelman to Stevenson, 20 January 1930.
55 I reached a similar conclusion in my exploration of public health in industrial Cape Breton. See Twohig, 'Public Health in Industrial Cape Breton, 1900–1930s.' In that case, the Sydney branch of the Victorian Order of Nurses considered the need to balance the number of Roman Catholic and Protestant nurses and raised the issue of employing local women over women who had specialty training in public health nursing.
56 CTA, File 48, Executive Secretary [R.E. Wodehouse] to Mr. V.H. Smith, Confederation Life Association, 14 July 1931.
57 See CTA, File 54, Creelman to Wodehouse, 29 April 1931 and Wodehouse to Creelman, 30 April 1931.
58 M. Sandelowski, *Devices and Desires* (Chapel Hill: University of North Carolina Press, 2000), p. 83.
59 P.L. Morrison, 'The Nurses in Hospital Administration,' *Canadian Nurse*, 36:10 (1940), 672–4.
60 A. Wright, 'Administration in Small Hospitals,' *Canadian Nurse*, 37:4 (1941), 230.
61 This advertisement appeared in *Canadian Nurse*, 36:12 (1940). Further evidence can be found in advertisements from either hospitals or individual workers, which often referred to multiple skill sets, including some combination of nurses' training and X-ray experience or training in X-ray and laboratory work. See advertisements in *Canadian Nurse*, 37:5 (1941), 358 and *Canadian Nurse*, 37:7 (1941), 493. Kathryn McPherson has ably noted how nurses were encouraged to undertake training in many facets of hospital work, including X-ray technique. See K. McPherson, *Bedside Matters: The transformation of Canadian nursing, 1900–1990* (Toronto: Oxford University Press, 1996), p. 221.
62 P.E. Hunt, 'Better X-Ray Diagnosis in Small Hospitals,' *Canadian Hospital*, 19:12 (1942), 48–9.
63 *Canadian Hospital*, 7:10 (1930), 36.
64 *Canadian Hospital*, 8:3 (1931), 42.
65 *Focal Spot*, 2:2 (1945), 98–9.
66 *Focal Spot*, 2:3 (1945), 104.
67 *Focal Spot*, 3:1 (1946), 52.
68 'News Notes,' *Canadian Nurse*, 45:12 (1949), 943.
69 Twohig, '"Local Girls" and "Lab Boys"'.
70 T.L. Adams, *A Dentist and a Gentleman: Gender and the rise of dentistry in Ontario* (Toronto: University of Toronto Press, 2000), pp. 110, 173.

71   Sandelowski, *Devices and Desires*, p. 3.
72   Sandelowski, *Devices and Desires*, pp. 63–4, 72.
73   Stevens, *In Sickness and in Wealth*, p. 12.
74   S. Reverby, '"Neither for the Drawing Room nor the Kitchen": Private Duty Nursing in Boston, 1873–1920,' in J. Walzer Leavitt and R.L. Numbers (eds), *Sickness and Health America: Readings in the history of medicine and public health* (Madison: University of Wisconsin Press, 1997), 253–65, p. 260 and Sandelowski, *Devices and Desires*, pp. 83–6.

Part II

Context, Contingency and the Life Stories of Technologies

6

Artificial Eyes and the Artificialization of the Human Face

Neil Handley

A facial prosthesis is an artefact, requiring art and skill to make.[1] It may be considered more humble and more functional than a fine oil painting but the technical accomplishment required to make it may be equal and the emotional effect it can have on both wearer and spectator may even be greater; to its owner a prosthesis is no mere material possession like a prized cooking pot or the last stamp required to fill an album.[2]

Historically, facial prostheses, (and specifically artificial eyes)[3] have been concerned more with the minimization and concealment of defects than they have with the replacement of function, though as this article will suggest, the supplementary benefits thereby accruing may impinge upon many other functions of everyday living.[4] As yet an artificial eye cannot help its owner to see, unlike 'false' teeth that may help him to chew or the artificial leg with which he may learn to walk. Where, however, there is a patient with a need for an artificial eye the provision of one, whether by a commercially motivated craftsman or a medical practitioner, can have profound effects.

A study of the artefacts themselves reveals much about the use of new materials, the occupations involved in the supply of prostheses, the extent of charity in a given period and the motivations for it. As a collector of such artefacts The British Optical Association Museum's interest is in prosthetic eyes as both technological and symbolic objects.[5] Although material culture studies on objects as diverse as shaman sticks or copper kettles have been around since the 1960s, this form of analysis has rarely been applied to medical prostheses, nor does this article attempt as much but the reader may at least see how approaching the subject from this angle has informed its conclusions.

It is important to remember that a prosthesis replaces an *organic* item – the original of which would never itself be considered an artefact, except as the handiwork of God. This may lead to a lack of respect for the replacement item, which can at best be regarded only as a substitute, no matter how expensive the cost of its production or complex the skills required in its fabrication.

Notions of physical beauty and what is 'normal' allow us to bastardize a recognized cultural history model and ascribe to the human face the role of a 'habitus' – a form of cultural capital invested in the bearer that can open doors to employment or social acceptance. Since Ancient times the bearer of the deformed or damaged face has encountered varying degrees of social and cultural discrimination and been denied the benefit that accrues from possessing a perfect body. Systems of punishment have sometimes recognized this, hence the deliberate mutilation of the ears, noses and eyes of criminals in many societies. The outward appearance of the major sensory organs has taken on a symbolic significance that sometimes goes beyond their functional importance. The pioneer of structuralist theory, Edward Leach, from Rochdale in Lancashire, wrote of the premise that all people are fascinated by bodies, particularly the orifices, of which the eye sockets (with their contents) number just two.[6] These orifices have had a symbolic religious significance in many cultures with notions of the 'all seeing eye'. An artificial eye fails to fit in completely with this structure since, however realistic it looks on first glance it can not hope to replicate the all important 'betwixt and between' – those boundary-crossing elements that so define the nature of the orifice such as tears, sleep deposits, blood, or falling eyelashes.

Historians could consider facial prostheses as they would clothes and costume or, for that matter, make-up and cosmetic scents. All these sets of objects illustrate the belief held at any one period in the value of perceived attractiveness (or at least, lack of repulsiveness) and indicate a strong motivation on the part of people to conform. Artificial eyes are not for seeing, but are for being seen, being located smack in the middle of the face – the most visible part of the human body. Early false teeth were also designed entirely for the sake of appearance; from the seventeenth-century Netherlands comes a description by Beverwijck of ivory teeth attached to the remaining natural teeth by means of gold wire, but these were quite unsuitable for biting or chewing. Tooth decay was presumably far more prevalent than eye loss so the lengths that people would go to conceal it are perhaps more surprising.

Some patients went beyond their desire to conform, choosing instead to make bold statements, drawing on the fact that the very inert materials most suitable for prosthetic use were precious metals, such as gold. The Danish astronomer Tycho Brahe lost the bridge of his nose in 1566 and required a partial prosthesis, which he made himself of gold and silver. Only a faint bump on his portraits gives a clue as to the presence of this. One account suggests that the precious metal was painted to blend in with the color of his skin. With this realistic substitute for a nose he was demonstrating the scientific achievement of the sixteenth century, most notably of himself, but we may also detect here the craftsman trespassing on God's own terrain in that he was making something that was more valuable than a real nose. He may not always have worn this *de luxe* proboscis. Traces of copper found in Brahe's coffin have caused at least one historian to speculate that he must have had a more humble example for 'every day use'.[7] Normally, of course, a prosthesis works best if it is not noticed. In this way it removes barriers to normal interaction and communication. One doesn't normally show off one's new prosthesis!

We can conceive of an artificial eye as a deliberate 'fake' object, intended to deceive.[8] In 1867 for example the doctor F.P. Ritterich examined a man recently returned from the Australian gold mines whose false right eye had been examined and allegedly declared 'normal' by four village physicians! As with the wig, the denture or camouflage make-up, all things with which the artificial eye shares a metonymic relationship, it is a form of a disguise that may enjoy varying degrees of success in its intended aim. To the student of material culture this causes a difficulty. If all prostheses are fakes, how can there be such a thing as a 'genuine' one? The truth – that is the deformity – which is what nature has crafted either at birth or through subsequent disease or accident, may be embarrassing, culturally unacceptable, unsightly, even hideous to behold. The viewer of the prosthesis, if he or she realizes it is such, may be left feeling indignant at the attempted deception (often the case with hairpieces) and may scorn the vanity of the wearer. On the other hand the viewer may be grateful. Few would prefer to gaze upon the British Chancellor Gordon Brown's left eye, lost through a rugby injury, if it were not cloaked with a prosthesis and the Austrian racing driver Nikki Lauda whose ear was mangled in a serious accident is both admired and yet sometimes criticized for his decision not to disguise the injured outer pinna. This raises the idea that today's availability of some forms of prosthesis on the National Health may be at least partly due to protecting the sensibilities of other people who might otherwise have to look at the disfigurement. Correcting the perceived problem can also be judged as indicative of a stable society. A 'social pathology' along the lines set out by Durkheim might support this further influence; a society providing prostheses is well adjusted internally.[9]

The artist or the social philanthropist has contributed as much as the medical practitioner in driving the necessary technology forward. Indeed the specialist technicians involved have, at times, been isolated by the medical profession – with its preference for pursuing the dream of full reconstructive surgery, something that was difficult but at least conceivable for the nose or the auricle of the ear, less so for the jelly-like eye.[10]

In the fifth century BC Roman and Egyptian priests made eyes from painted clay, which were attached to a cloth and worn over and outside the socket. This type is known in Greek as the *ekblepharon*. The same artistic skills were also applied to death masks around 2000 BC. The eyes on those masks found associated with Egyptian mummies, often made of plaster-filled bronze, resemble cosmetic replacements. Roman statues were also embellished with semi-precious stones in the eye sockets. Makers of these eyes were called *faber ocularius*, which is distinct from the *medicus ocularius*. The work of Ambroise Paré from 1579 (published posthumously in 1614) features illustrations of artificial noses held on with strings and artificial eyes on the end of curling rods that bend round the head. In 1561 Paré had been the first man we know of to recommend placing a gold and colored enamel artificial eye, into the actual orbit of a living patient – the type known as *hypoblephara*; this seems like an incredibly late historical development but then Paré did not claim that the idea was new.[11] In a British private collection I have seen possibly the oldest *hypoblepharon* still in existence. It is made of wood and ivory and certainly looks the part but lacks provenance. There is much research still to complete, including an investigation into those facial prostheses (such as there are) preserved in public museums. We can observe that there was a parallel development in the sixteenth century of artificial noses and ears, spurred mainly by the injuries incurred when fighting a duel – the very cause of Tycho Brahe's disfigurement mentioned earlier.

In the second half of the sixteenth century and for two centuries thereafter the Venetians made glass eyes by a secret method, apparently with fragile, crude and uncomfortable results. To begin with these workers were craftsmen, skilled glassblowers and lens makers, controlled by restrictive guilds and rarely permitted to travel. Glass eyes seem also to have come from Augsburg, another historic centre of the optical manufacturing industry, until the devastation of the Thirty Years' War benefited the French whose makers were mainly enamellers rather than glassblowers.[12] English practitioners were few and far between. In September 1679 William Boyse of London advertised artificial eyes in the *True Domestick Intelligence* and informed readers that they must have known him for many years as '*the only person* [my italics] expert in making artificial eyes of enamel, coloured after nature... in which not only fitted the socket with ease to the wearer, but turned with all the facility of the real organ of vision'. His advertisement of 1681 in *Merlin's Ephemeris* proclaimed him 'the only English operator in glass and the most expert in making artificial eyes so exact as not to be distinguished from (the) Natural, they are of enamel, with colour mixt of the same, without either paint or lead, and worn with much ease, and so curious that they have the motion of the natural eye, being exactly made to the colour or bigness of the same which renders them very ornamental and commodious, the like was never made in England'.[13]

William Boyse was succeeded, in around 1710, by his son-in-law, James Smith, of whom a famous portrait print survives, clutching an exquisite example of his wares.[14] A text published in London in 1699, apparently a translation of Henri de Blancourt and claiming to reveal a secret method of manufacture to the public recounts how a crystal eye could be blown, complete with imitation blood vessels.[15] In the eighteenth century there was a brief period of Parisian dominance in the art of artificial eye-making. Glass, enamel and metal eyes continued to be available. In 1752 Heister, a surgeon of Nuremberg wrote that he preferred glass eyes to ones made by goldsmiths since metal ones lost their brightness and repelled tear fluid. Johann Friedrich Dieffenbach was a School of Berlin Surgeon from 1823 who set himself up as a champion of the poor. Children chanted in the street: 'Who does not know Dr Dieffenbach? / The doctor's doctor! / He amputates arms and legs and *makes new* [my italics] noses and ears.'[16] This last part of the song was in fact a reference to his plastic surgery but the same philanthropic motivation is to be observed in his compatriot Friedrich Philipp Ritterich of the University of Leipzig – where he founded the institute for poor eye patients. The role of Ritterich (1787–1866), an individual doctor and teacher, in sponsoring the growth of a nineteenth century glass-blowing industry in Germany cannot be underestimated.[17] In 1852 Ritterich wrote: *Das Künstliche Auge*, a monograph that confirms his concern for his fellow man.[18] From 1810 and for 30 years he had fitted artificial eyes as part of his general medical practice. Twice he had gone in person to buy stock eyes in Paris. Otherwise he had 400–500 samples sent to him at his office so he could choose the best one. These he then provided to his patients at cost (5 or 6 Thaler). This was in stark contrast to Boissoneau in Paris who demanded 25 Louis d'Or and said 'the social status of the patient should be noted because the fee will be determined by the economic condition of the patient. Indigent patients will be charged 25 Francs whereas for a rich patient the sky may be the limit'.[19] In order to avoid this expensive importation Ritterich encouraged German glassblowers and even organized the training of craftsmen in this task. He persuaded Leipzig Eye Institute to provide eyes free of charge to the indigent on the grounds that obtaining an artificial eye was more important for the poor than for the rich; it protected the poor against misery and from need.[20]

We can trace the gradual journey of artificial eyes from a commodity, supplied from stock, to an individually crafted device, provided by a compassionate society.[21] Boissoneau could supply symmetrical eyes that were supposedly suitable for insertion into either socket. As we have seen, by the mid-nineteenth century Germany was dominant. In 1832 the glassblower Ludwig Müller-Uri developed a glass eye at the famous Lauscha Glass Factory (est. 1597). German craftsmen began to tour the USA stopping at cities for a few days to demonstrate glass techniques. The Americans started making their own eyes in about 1850, with Pierre Gougelmann, who had trained under Boissoneau, being an early pioneer although German products continued to be considered the best – some of those from the village of Lauscha were produced by successive generations of a few families working from a cottage industry.[22] Glass eyes illustrating diseases were used as teaching aids. An Austrian hand-blown set was used by the San Francisco Eye Institute (Greens Eye Hospital) from 1870. The British Optical Association Museum in London possesses another such collection including eyes made for abnormal cases such as contracted sockets, specimens with special prongs and spurs, ptosis crutches and partial lids or other artificial surrounding tissue.[23]

In London the commodity was produced in much the same area as the spectacle frame and lens industry as a search of Kelly's Directory proves.[24] Addresses such as the Goswell Road, Smithfield and Fleet Street occur in connection with artificial-eye makers from the 1830s onwards. In 1840 James Gray of Shoreditch is listed as combining eye-making with the manufacture of mineral teeth, whilst the Arnolds of West Smithfield were principally surgical-instrument makers. By 1870 Masters Moses of the New Kent Road could boast of two medals awarded (in London and Paris) for their artificial hands, legs and eyes.

The range of eyes available in Edwardian England was recalled by Mr A. Bernard Clark (of Dollond & Co) whose 1950s memoirs record that: '*Most* [my italics] opticians stocked artificial eyes. They were stocked in trays holding a hundred or more', but these were '*for display purposes chiefly* [my italics]. Mostly the eyes actually issued to patients were made to special instructions, for example a custom-made Snellen eye used to cost the patient a guinea including fitting'.[25]

In 1900 Pache & Son of Birmingham were makers to the principal hospitals in the United Kingdom and could provide Snellen's improved 'reform eye' – a double shell that prevented a sunken appearance.[26] In fact the closed eye, in which the cavity was filled with wax to give it greater bulk though, it seems, little additional weight, was developed almost concurrently in two places.[27] J.L. Borsch, a Philadelphia doctor, claimed to have visited Herr Müller in Wiesbaden in 1894 and commissioned from him such a prosthesis. These were subsequently tried at the de Wecker clinic in Paris and a presentation given by Borsch's colleague, Dr Schwenk, back in Philadelphia in 1897, one year before Snellen presented the 'reform-auge' in Rotterdam, he also having called upon Müller to manufacture the item.[28] Despite this rare instance where an ophthalmologist, whether Snellen or Borsch, had been instrumental in a radical design innovation, the driving force behind the invention was directly related to a development in surgical practice: Enucleation of the orbit was now more common, whereas previously most ocular prostheses had been intended to cover an atrophied eye. Hazard-Mirault had written back in 1818 that many ophthalmologists were simply unwilling to risk the operation that would be necessary to prepare a socket merely to facilitate the insertion of a prosthesis.[29] Such surgeons thought, perhaps with some cause, that they were sparing their patients misery. Later generations of the Boissoneau family continued to call themselves *ocularistes*, (eye makers) to distinguish themselves from the *oculistes*, (eye doctors) with whom they seldom communicated. A century on a scan of the optical journals reveals that in 1927 Rose Millauro, of London was advertising that 'Every type of Artificial Eye embodying all modern improvements in shape, finish and aesthetic, can be supplied by the actual manufacturer .... The existence of a good eye-maker should be of great interest to the Ophthalmic Surgeon; bad artificial eyes invariably spoil the result of a successful operation'. That is to say that Millauro was making the point that however disinterested they might otherwise be, senior surgical practitioners had a self-interest in ensuring that the work of the technician was of the highest order![30]

Also in 1927, Gustav Taylor, Artificial Eye Maker, advertised 'The Sympathetic Artificial Eye – The Most Perfect Eye Invented':

> [Mr Taylor] who has made Artificial Eyes for twenty-eight years, has invented an undetectable artificial eye which moves and acts in sympathy with the natural eye. He has also invented eyes with dilating pupil. These eyes are the most lifelike and natural ever made, and a perfect likeness to the other eye.... All eyes made on the premises in front of the patient within an hour by G. Taylor personally. Doctors are especially invited to see the maker at work at the above address.[31]

We may note especially Taylor's use of the word 'invented'. It harks back to my point about the prosthetist replacing the natural with a new piece of technology devised by man in a manner akin to the steam train, the telephone and the light bulb.[32] I note also the assumption that the all-important doctors would travel to see him. Would they in fact have bothered? The role of the optician as supplier, salesman and occasionally craftsman can be contrasted with that of the ophthalmologist who would often discharge a patient following a successful enucleation.[33] A 1935 article in *The Optician* by F. Kemble Williams suggests a lack of doctorly concern.[34] He writes that no one else but the optician seems prepared to undertake this work of selling eyes. Even the eye surgeons appear to think that it is part of the optician's job and always refer their patients to an optician after the enucleation. Despite this there is no mention of it in the exam syllabus of either profession. The provision of prosthetic eyes is still regarded then as a commercial transaction rather than a medical service. Williams claims that most opticians do it occasionally but many neglect basic hygiene – which is most unwise as the patient may be dirty! He also complains that manufacturers have failed to provide an accurate method of ordering the eyes and that numbered patterns and color charts would be useful. It could, however, be questioned whether such a scientific and 'professional' method was justified by the small number of patients. Around 1946 the British firm of W.H. Shakespeare & Sons Ltd could claim that its 36-piece fitting set was sufficient to meet the needs of over 80 percent of patients. The practitioner had simply to select a numbered shape, a numbered iris color and sclerotic and issue any additional instructions such as moving the position of the iris.[35]

One benefit of eyes supplied from stock, that still applies today, is that they can be good for children who will undergo normal growth (tissue change within the orbit) thereby not wasting money spent on custom-made prostheses. Nevertheless developments in material science have made the custom-made eye the more prevalent. A crucial development was the introduction of better materials for taking impressions of the eye or its empty socket. Unhappy with the usual impression material of the time, Negocol, the dispensing optician and instrument designer C.H. Keeler (1903–1993) persuaded the Amalgamated Dental Company to produce a better material. This was introduced as Ophthalmic Zelex[36] by his company, C. Davis Keeler, and was supplied by two other major players in the ophthalmic goods business, Clement Clarke Ltd and Theodore Hamblin Ltd.

Changes in glass-making technology and the invention of new acrylics have been crucial in affecting the practices and training of the modern ocularist. During World War II a shortage of German (Lauscha) glass led the British and the Americans to investigate techniques for the use of acrylic. One notes in particular the US Government's involvement in partnership with manufacturing firms including one led by Paul Gougelmann. Research was carried on in parallel on both sides of the Atlantic. Royal Navy dental technicians were probably the first to use plastic in 1941. Meanwhile Fritz W. Jardon (in Southfield, MI) in conjunction with the American Optical Co and the US Army and Navy Dentist perfected Methyl-Methacrylate resin.[37]

Acrylic became the dominant material to the present day. It had durability and longevity. It was more comfortable for many patients. It did not allow fluids to build up behind the prosthesis and could be worn in bed. This led to a high public demand that eventually outweighed the preferences of many artificial-eye fitters for glass, not least because their highly developed glass blowing skills made them reluctant to make the switch. Shakespeare & Sons issued a promotional leaflet including a discussion of the 'Perfect Plastic (acrylic) Eye', (Patent No. 580, 993 – dated 1946) and its advantages over glass. In particular a plastic eye held in stock by the optician could be ground to size, making it a better business proposition. Nevertheless acrylic eyes required re-polishing every eight months to prevent infection – the drug treatments for which, if needed, would damage the prosthesis. In Europe, therefore, glass eyes did continue to remain available using Cryolite glass on account of its many advantages, including its natural appearance, compatibility to match the shading of skin tones, hard surface and scratch resistance.[38] As a hygroscopic material it retained natural eye fluids (unlike acrylic which is a water-repellent material) and could also be cleaned hygienically with just cold water.

It is crucial to point out that this is not one long success story: artificial eyes have frequently failed in their intended aim of concealing the loss of a living eye, hence the popular music hall song:

He's very well known is Algy
As the Piccadilly Johnny with the little glass eye

At least the song speaks of Algy in endearing terms. Maybe he would have been proud of his eye, removing it almost as a party-piece. Artificial eyes have proved uncomfortable to wear; some have reacted with bodily fluids; others have actually imploded because blown glass shells contained internal vacuums! This was the case in 1916 when *The Ophthalmological Record* mentions several cases of shattering eyes.[39] Four patients suffered this unnerving experience on two separate occasions! There was also little that could be done for patients who had lost the surrounding socket though the French Dr Henri Einius produced a prosthesis in 1917 that could be worn even with no remaining eyelid. The driving force of war as a vehicle for technological change is evid-

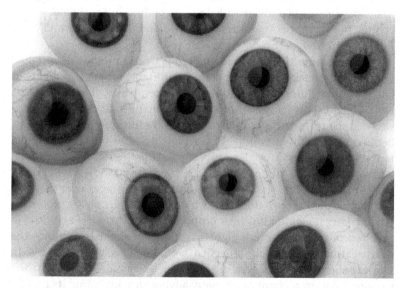

Figure 6.1 Artificial eyes from an Army Spectacle Depot. In the early twentieth century opticians believed they could supply any shape, size or color of artificial eye from stock. These eyes in the British Optical Association Museum show a representative range, indicative of the manufacturer's skill. Photograph reproduced with permission of The College of Optometrists.

ent here.[40] Indeed the Army Spectacle Depot established by John Sutcliffe at the British Optical Association during World War I was originally intended to supply spectacles to conscripts, the eyesight requirements for servicemen having been relaxed due to manpower shortages. The Depot soon found itself supplying ocular prostheses for the injured returning from the trenches and this formed the basis of the National Artificial Eye Service (NAES) still in existence today and based in Blackpool. A standard Army Spectacle Depot fitting set contained 150 different eyes.[41]

Better examples of artificial eyes were designed in the twentieth century due to a switch from artificial eyes being considered a cosmetic remedy (allowing you to discard the bandage or eye patch) to a *hygienic* remedy! Greater awareness arose of the after-care needs of prosthesis wearers for whom the underlying problems might continue to include dry eye, recurrent growths and infection.

Modern examples of artificial eyes now interact with living tissue. Since the early 1980s, when Dr Arthur Perry developed an extraordinarily bio-compatible implant made of marine coral, doctors have been able to provide a relatively complication-free, reliable way to achieve remarkably lifelike movement.[42] The use of a form of calcium carbonate, hydroxyapatite, derived from natural sea coral has also proved revolutionary as it is very porous and blood vessels can grow over it. As the body absorbs the material it ceases in a way to be artificial.

The great prize throughout the history of artificial eyes has been the pursuit of replacement eyes that actually see, however this dream has for most of history seemed so unrealistic as not to be pursued seriously. Early efforts in the second half of the twentieth century involved head-mounted camera-type contraptions that would have had serious cosmetic drawbacks whatever their functional benefits. They did not deflect attention! In March 1994 one respected British journal could declare: 'artificial eyes – not something contemporary ophthalmologists spend much time on'[43] and the artificial-eye maker was still considered a somewhat peculiar character as evidenced by the portrayal of Hannibal Chew, creator of Nexus 6 Replicant eyes in the cult film *Blade Runner* (1982). This situation started to change, albeit slowly. Artificial-retina development was pioneered at the Johns Hopkins University in the United States. In 1994 surgeons implanted an electrode to the eye of a blind patient who claimed to be able to see a black dot surrounded by a yellow ring. In the Spring of 1997 a professor of electrical engineering, Wentai Liu, created a microchip for surgical use that had an array of pixels sufficient to identify individual letters. Concurrent work on bio-compatibility at Stanford University resulted in a new synthetic cell membrane that could adhere to living cells and silicon chips. An ocular prosthesis with a 'bionic eyelid' made of elastic latex was developed at Humboldt University, Berlin in 1999. Ongoing research throughout the 1990s into replicating the function of the retina led to the announcement in 2003 by the Association for Research in Vision and Ophthalmology in Fort Lauderdale that three US patients had been implanted with a sliver of silicone and platinum studded with 16 electrodes which stimulated remaining healthy retinal cells and this passed visual information to the optic nerve.[44] I conclude from this growing area of interest that there is consequently even a risk of the term 'artificial eye' changing its common meaning and becoming the concern, if not the preserve, of the ophthalmologist after all.

Acknowledgments

The author is grateful to Mr Colin Haylock of the Charing Cross Hospital , Hammersmith, Mr Robin Brammar, Oculist, and to Mr Tim Bowden of the City & Islington College for their assistance in the research for the original paper upon which this article is based.

The vast majority of the works mentioned in the endnotes, together with various actual examples of ocular prostheses are available in the British Optical Association Museum and Library at the College of Optometrists, 42 Craven Street, London WC2N 5NG. Contact either: library@college-optometrists.org or museum@college-optometrists.org and see the 'MusEYEum' website at http://www.college-optometrists.org/museum.

Notes

1 See: M.O. Jones, *The Handmade Object and its Maker* (Berkeley: University of California Press, 1975): 'An object cannot be fully understood or appreciated without knowledge of the man who made it'.

2 See especially: A.H. Degenhardt, 'Artificial Eyes,' *Vision*, 4:3 (1950), pp. 12–16, (especially page 12 where it is stated 'their purpose...is primarily that of increasing the happiness of a person who would otherwise be subject to a chronic self-consciousness of his misfortune'); V.A. Chalian *et al*, *Maxillo-facial Prosthetics* (Baltimore: Williams and Wilkins, 1971); D.J. Reisberg and S.W. Habakuk, 'A History of Facial and Ocular Prosthetics,' *Adv Ophthalmic Plast Reconstr Surg*, 8 (1990), 11–23 (especially page 11) and K.F. Thomas, *Prosthetic Rehabilitation* (London: Quintessence Books, 1993).

3 The term 'artificial eye' can be misleading since, in optics, it is also used to refer to instruments that imitate the inner workings of the eye so as to facilitate the study of image formation and clinical training in the use of the ophthalmoscope and retinoscope. One well-known American instrument

supplier marketed the two types in adjacent sections of its catalogue: James W. Queen & Co, 'Artificial Eyes', in: *Priced and Illustrated Catalogue of Ophthalmological Instruments, Spectacles, and Eye-Glasses* (Philadelphia, 1889).

4   This article and the original conference lecture on which it is based actually grew out of the British Optical Association Museum's research into the history of contact lenses – a form of optical device that can have both a cosmetic and a corrective function. The history of contact lenses and artificial eyes is closely intertwined and yet separate. Ophthalmic and Dispensing Opticians have been influential in driving forward both technologies, but few have dealt in both at the same time.

5   See also: R. Coulomb, *L'Oeil Artificiel* (Paris: Ballière et fils, 1905); M.O. Hughes, *Eye Making: A brief history of artificial eyes made in Virginia, Washington, D.C. and Surrounding Areas* (Vienna, Virginia: Artificial Eye Clinic, 2000); S. Marinozzi, 'Eye Surgery from the origins to Antonio Scarpa,' in: R. Riccini, *Taking Eyeglasses Seriously. Art, history, science and technologies of the vision* (Milan: Silvana Editoriale, 2002), 151–61; R.W. Nowakowski, *Primary Low Vision Care* (Norwalk: Appleton & Lange, 1994); P. Pansier, *Traité de L'Oeil Artificiel* (Paris: A. Maloine, 1895); A. Poli, 'Augmented Vision,' in: Riccini, *Taking Eyeglasses Seriously*, 179–81.

6   E.R. Leach, *Culture and Communication: The logic by which symbols are connected* (Cambridge: Cambridge University Press, 1976).

7   V.E. Thoren, *The Lord of Uraniborg: A biography of Tycho Brahe* (Cambridge: Cambridge University Press, 1990).

8   For the theoretical framework, see M. Jones, *Fake? The art of deception* (London: British Museum Publications, 1990).

9   Hodder hints at this in a different context. See I. Hodder, *Theoretical Archaeology: A reactionary view* (Cambridge: Cambridge University Press, 1982). To consider the relevance of Durkheim see: A.R. Mawson, 'Durkheim and Contemporary Social Pathology', *Br J Sociol*, 21:3 (1970), 298–313.

10 For example see: M.I. Berson, 'Complete Reconstruction of Auricle,' *Am J Surg*, 60 (1943), 101–4.

11 J. Hirschberg, *A History of Ophthalmology*, Volume 4, (Bonn: Wayenborgh, 1984), 33–7. Hirschberg also demonstrates the erroneous nature of claims that artificial eyes are mentioned in the Jewish Talmud.

12 Anon, 'Artificial Eye Making Abroad,' *Optician and Scientific Instrument Maker*, 41 (1911), 205–7.

13 Quoted in: Anon, 'Notes and Echoes: Glass eyes in 1681,' *Br J Ophthalmol*, 22 (1938), 445.

14 *James Smith Artificial Eye Maker at the Brunswick Coffee House in Fleet-Street*, engraved by John Pine after an original portrait by F. Tunks (1717). BOA Museum object accession number LDBOA1999.149 purchased by the museum in 1958.

15 C.J. Robb, 'Artificial Eye Making in 1699,' *Optician*, 144:3726 (1962), 197.

16 Quoted in J. Hirschberg, *A History of Ophthalmology*, Volume 5 (Bonn: Wayenborgh, 1985), p. 303.

17 For a summary of Ritterich's career in Leipzig see J. Hirschberg, *A History of Ophthalmology*, Volume 6 (Bonn: Wayenborgh, 1986), pp. 183–96.

18 F.P. Ritterich, *Das Künstliche Auge* (Leipzig, 1852).

19 Quoted in Hirschberg, *A History of Ophthalmology*, Volume 6, p. 188.

20   This is Hirschberg's summary paraphrase of Ritterich's writing on the subject, quoted *ibid.*

21   See A. Appadurai, 'Commodities and the Politics of Value,' in: A. Appadurai (ed.), *The Social Life of Things* (Cambridge: Cambridge University Press, 1986).

22   Artificial-eye making remained a family concern, often somewhat secretive in its methods, in much of France and Germany until the early twentieth century. See: G.J. Jones, 'Artificial Eyes,' *Dioptric Review*, 38:4 (1936), 187–94.

23   Anon, 'The B.O.A. Collection of Eye Disease Models,' *Dioptric Review*, 39:5 (1937), 400.

24   *English Eye Makers As Listed in Kelly's Post Office Directories for London.* Unpublished document in BOA Museum files.

25   Quoted in B.S. Ibbs, 'The Cradle of British Optics Part VI,' *Optician*, 134:3483 (1958), 643–4.

26   The famous Royal London Ophthalmic Hospital at Moorfields obtained its artificial eyes from Gray and Holford of Goswell Road. On Pache & Son see C.H. Lea, 'Artificial Eyes,' *The National Optical Journal* (1941), 102.

27   J.L. Borsch, 'The Closed Artificial Eye,' *Dioptric Review*, 5:46 (1901), 271; I. Den Tonkelaar, H.E. Henkes and G.K. van Leersum, 'Herman Snellen (1834–1908) and Muller's "reform-auge." A short history of the artificial eye,' *Doc Ophthalmol*, 77:4 (1991), 349–54.

28   Borsch, 'The Closed Artificial Eye'.

29   C.F. Hazard-Mirault, *Traité Pratique de L'Oeil Artificiel* (Paris, 1818).

30   Rose Millauro was a doctor's daughter who set up in practice at 30 Fitzroy Square, London. In 1934 she commenced fitting artificial eyes even to young babies at the request of the National Institute for the Blind. She was not the only woman to so practice. Earlier, in 1894, Miss Priscilla Shaw of Counden Street, Coventry patented the adaptation of an artificial eye in the form of shell attached to a spectacle frame.

31   From an advertisement in the BOA Museum subject files.

32   Taylor took out several British patents including 175212 (the one alluded to above) for a pupil that appeared to dilate under different lighting conditions, 175213 for the use of acetyl acetate or shellac to coat the glass eye and reduce its fragility and 177074 a modification of the 'reform' eye to improve its comfort during wear.

33   For Trousseau's criticism of this habit see: J.H. Sutcliffe, 'The Care of Artificial Human Eyes,' *Dioptric Review*, 2:6 (1897), 58–61. As late as the 1950s the acknowledged British expert on artificial eyes was an ophthalmic optician, J.H. Prince FBOA, not an ophthalmologist. Prince had travelled extensively across the world and regretted that the submergence, as he saw it, of the British 'craft' into a 'Social Service Scheme' would bring an end to its international dominance in quality of workmanship. See: J.H. Prince, *Recent Advances In Ocular Prosthesis* (Edinburgh: E&S Livingstone, 1950).

34   F. Kemble Williams, 'The Optician and Artificial Eyes,' *Optician*, 89:2294 (1935), 83.

35   Anon, *Plastic Artificial Eyes* (Leigh-on-Sea: W.H. Shakespeare & Sons, undated, probably 1946).

36   T. Bowden and A. Gasson, *Contact Lens Pioneers* (London: BCLA, 2004), p. 8. 'Zelex' was the original alginate impression material, though it was not

without its own problems. It was important to use tepid water as cold water might retard the setting of the material.

37 P.J. Murphey, L Schlossberg and L.W. Harris, 'Eye Replacement by Acrylic Maxillofacial Prosthesis,' *Naval Med Bull*, 43 (1944), 1085.

38 O. Martin and L. Clodius, 'The History of the Artificial Eye,' *Annals of Plastic Surgery*, 3:2 (1979), 168–70.

39 J. Jameson Evans, 'Rochester, Alexander S. (Chicago) – The spontaneous explosion of Snellen improved artificial eyes,' *Ophthal. Record*, December 1916, reviewed in: *Br J Ophthalmol*, 5 (1921), 522–3.

40 Anon, 'A New Artificial Eye,' *Optician and Scientific Instrument Maker*, 53:1361 (1917), 111–12.

41 For example see BOA Museum Object Accession Number 2005.373 donated by the NAES. This is a glass-topped wooden box with 150 internal compartments each containing an artificial eye in varying shapes, sizes and colors.

42 See A.C. Perry, 'Integrated Orbital Implants,' in S.L. Bosniak and B.C. Smith (eds), *Advances in Ophthalmic, Plastic, and Reconstructive Surgery*, Volume 8. (New York: Pergamon Press 1990), 75.

43 F. Roman, 'The History of Artificial Eyes,' *Br J Ophthalmol*, 78:3 (1994), 222.

44 Anon, 'Bionic Eye,' *Optician*, 225:5902 (2003), 5.

7
Biotronik: 40 Years of German Entrepreneurship in Medical Technology

Patrik Hidefjäll

Introduction: the cardiac rhythm management device industry

The first fully implantable pacemaker was used clinically to save the life of a patient in Stockholm in 1958. Soon thereafter many research teams in what was then known as the 'medical electronics' field developed their own versions of cardiac pacemakers, thereby collectively constituting what would later be termed the cardiac rhythm management (CRM) device industry. Biotronik, based in West Berlin, joined this industry in 1963 as a start-up company with its own pacemaker. Many other companies have entered and left the CRM industry since the early-1960s. Today four companies dominate the industry with more than 97 percent of worldwide revenues; all of them except Biotronik are US companies. Biotronik and the industry leader Medtronic are the senior members of this small group.[1] In this paper I will examine Biotronik's role in the cardiac rhythm management industry, and explain how it, as the only European company, was able to maintain its position in the face of competition by larger, publicly listed US medical device companies. The paper will delineate the changing nature of the medical device industry and seek to understand the ways in which these changes affected the industry as a whole and Biotronik in particular.

These firms still compete fiercely to hold and expand their shares of the global market for implantable defibrillators, pacemakers, and related products. Much of this competitive striving is centred on corporate product development laboratories: medical device (including CRM devices) and pharmaceuticals at present are the most R&D (Research and Development) intensive of all industrial sectors. In the US, device manufacturers spent some $29,095 per employee or 11.8 percent of sales revenues on R&D in 2000, nearly twice the percentage for the next-highest sector, information and electronics.[2] New firms aspiring to enter the CRM industry face a formidable set of barriers including the wall of existing patents, the complexities of regulation, and long-standing relationships between company sales representatives and physicians. Even if small entrepreneurial companies develop promising new technologies, the established firms often acquire them before their products reach the market.

Implantable pacemakers have been commercially available since 1961, implantable defibrillators since 1985, yet the identity of both product classes continues to be redefined. Unlike pharmaceutical companies with their blockbuster drugs, companies have managed to update their CRM devices repeatedly by downsizing them, adding longevity, and expanding device capabilities with integrated microprocessors, sensors and lately telecommunication capabilities. Despite this, CRM devices of different manufacturers may not appear to differ greatly. The performance characteristics may vary in some degree, or one may boast a distinctive new feature that the other lacks. A manufacturer can gain market share, retain some pricing power, and add to its patent portfolio by introducing a succession of incremental changes in an established product. No company possesses all the technical knowledge required to produce state-of-the-art implantable devices; instead, companies have found that they must cross-license their patent portfolios. Incremental changes have yielded a stream of new patents that manufacturers can use as bargaining chips with other firms.[3]

A highly regulated industry

When the German medical device company Biotronik was founded in 1963, formal regulations of medical device use were minimal. Regulation of clinical practice occurred primarily by adhering to norms developed by the medical profession itself.[4] It was also within the medical profession that ideas for better devices were born and later tested. Emil Bücherl, the Berlin surgeon who picked up the idea to develop a German pacemaker while working with the Swedish physician Åke Senning in Stockholm, therefore had a crucial role in convincing his German peers of the efficacy and safety of the Biotronik device. But, with time other factors than professional norms gained more influence over medical device innovation in the CRM industry.

In addition to normative institutions, coercive institutions in the form of governmental market approval were introduced, exemplified by the US Medical Device Amendments of 1976. With world markets developing, shared understandings about what to expect of a pacemaker developed and were conceptually manifested as in the so-called Pacemaker Code, which initially was a three-letter code to signify the functionality of a pacemaker model.

Eventually, the cardiac rhythm management industry became highly influenced by normative, coercive and cognitive institutions, together forming an organizational field consisting of three spheres: the professional, the market and the societal sphere.[5] Normative institutions with their locus in the academia and the related professions provided norms for the evaluation of pacemaker technology and its use, coercive institutions in the form of legal requirements stated obligations of industry participants and regulated conflicts and resource flows to and within the industry. Within markets, shared perceptions of what to expect of products and services emerged. This increasing institutional complexity made the innovation process increasingly difficult for individual inventors as they had to comply with all these institutional structures to gain legitimacy and approval from several stakeholders for the realization of a new innovation.

Max Schaldach, the founder of Biotronik, liked to act on perceived opportunities which others had not yet noticed. He liked to challenge the given order and did not want himself or his company to become constrained by institutions. Like many other company founders he most of all valued and strived for independence. Even if an entrepreneurial drive for independence is an important prerequisite for founding a new business, or carry through a radical innovation, it may not be sufficient to master a subsequent competitive and increasingly institutionalized environment to stay ahead.[6] In such environments innovation is subject to structural forces of regulative, normative and cultural-cognitive character, institutional forces that need to be understood and adhered to, to survive in the long-run.[7] To increase the chances of success, an innovation needs to address a pressing need, a bottle-neck to further progress, but must also conform to taken-for-granted and therefore expected performance attributes. Even if innovation and entrepreneurship challenge given institutions in one respect, they still are subject to institutional forces in other respects.[8] Therefore, an entrepreneur needs to actively handle the institutions of innovation, the structures that support and select innovations, to successfully manage a continued stream of innovation.[9]

The ability to handle the institutions of innovation mainly stems from a management capability to integrate several diverse requirements into the innovations to be developed and commercialized. It requires the ability to integrate new ideas from company internal as well as external innovative individuals, at the same time considering the needs of advanced customers, position the company so as to attract venture capital and build the structures and processes to industrially exploit innovation. Most of all, the management of continuous innovation requires reading, responding to and shaping the expectations of customers and other stakeholders to manage today's dynamic business that has become fueled by an almost self-generating innovation system.[10]

Continuous innovation and the capabilities of firms

The popular desire for a steady stream of breakthrough products in the 'wars' on degenerative diseases often puts a spotlight on the device industry. In the United States, industry analysts and the news media take note of new research findings, new treatments and technologies.[11] One might assume that competition among companies in the CRM industry must centrally involve product innovation. However, the situation is more complex than that. If we ask what assets and capabilities are required for success under the conditions of intense competition and continuous innovation in this industry, we discover that the ability to design and build innovative products does not by itself guarantee survival with a respectable market share. Numerous medical device manufacturing companies have pioneered innovative products embodying striking technological improvements that competitors quickly embraced, yet the pioneering firms have often disappeared. Clearly, more was required than simply being good at technology.[12]

Consider the familiar definition of innovation as 'the initial commercial application of a new product or process.' This definition itself implies that innovation must involve not only research and product development but a company's capabilities in manufacturing, marketing, sales, and customer support.[13] Richard R. Nelson, following several earlier theorists of the business firm, has pointed out that 'firms have (at best) a limited number of things they can do well,' a set of 'core capabilities that enable them to be effective in the context in which they operate.'[14] Core capabilities, by definition, cannot consist of knowledge readily available to any interested party but must be out of the ordinary and difficult for competitors to replicate. For a device manufacturer a short list would probably include intellectual property accumulated painstakingly over many years (including the know-how to integrate several technologies into reliable, life-supporting systems), specialized plants and equipment, and manufacturing know-how.[15]

Companies may also come to possess complementary assets; these are assets perhaps less difficult to emulate but still important in ensuring that their products will be rapidly absorbed into the marketplace. David Teece, in an influential discussion on 'how to profit from innovation' distinguished between generic, specialized and co-specialized assets and mentioned marketing, manufacturing, after-sales or in case of systemic innovation other parts of the system;[16] one could add the company's existing customer base, its informal personal networks and reputation in the eyes of relevant audiences, and in our case its ability to carry out clinical trials and influence reimbursement rates. A company's complementary assets thus significantly influence the ability to commercially exploit an innovation. Due to the diverse and systemic technologies used in the CRM device industry and the specific requirements of customers a broad array of highly specialized complementary assets are needed. To develop all such complementary assets internally would be very costly, thus a tough challenge to a smaller privately-owned company such as Biotronik.

The co-evolution of companies and their environments

Every company must learn to do certain things highly effectively, but each will develop its own particular approach to these tasks; yet no company constructs itself entirely without external forces. The nature of the market and the governing institutions of a society, both general and specific to this economic sector, will influence the developing industry and individual firms. The environment in the US, with its high prices and readiness to reimburse new technologies fully once convincing scientific proof can be presented, has influenced the industry. At the same time, it is an environment very much influenced and dominated by the big US medical device companies, making it extraordinary difficult for non-US companies to participate. As a result, institutions have changed in response to activities in the industry and its growing sense of collective interest. Trade associations have formed, lobbying goes on, and public support has taken shape.[17] The different national healthcare regimes with which the companies became familiar in their early years and, more broadly, the role of unique historical events and choices have also oriented the companies toward certain future paths of development. Considering that the US market constitutes more than two-thirds of the worldwide cardiac rhythm management industry, it is of interest to ask how a non-US company such as Biotronik with a low US market share has survived in this very competitive industry.

Biotronik – a university spin-off

When the physics student Max Schaldach in the early-1960s was asked by the thoracic surgeon Emil Bücherl to develop a cardiac pacemaker, the goal was to create a German device of better performance at lower cost than other available pacemakers.[18] The project involved extensive reverse engineering, not only analyzing the clinical performance of competitive products, but also approaching other manufacturers such as Siemens-Elema and Cordis to learn as much as possible.[19] In this process, Max Schaldach also involved his student friend, the electrical engineer Otto Franke, to perfect the electronic circuitry; the two founded Biotronik in 1963. Biotronik was as much an academic as a commercial venture: the founders aimed to apply interdisciplinary scientific knowledge and engineering to the advancement of medical practice. Although Max Schaldach was an entrepreneur with commercial interests, his main desire was to gain recognition within academic communities engaged with medical technology and medical practice. Schaldach would be the one to imprint his personality on Biotronik through his nearly 40 years of leadership as owner-manager of the company.[20]

The growing interest and demand for the scientifically engineered Biotronik pacemaker allowed Otto Franke in 1965 to start working full time in the production of cardiac pacemakers, partly at his University institute, partly at home. In order to focus on the technological and scientific aspects of the pacemaker business, sales activities were contracted out to independent agents. This was an extremely innovative period, in which new improved prototypes, such as a metallic-encased pacemaker, a single-lead VAT pacemaker,[21] pacemaker stimulation for tachycardia, and biogalvanic energy sources were developed in close cooperation with Dr Bücherl.

However, Dr Bücherl and his assistants did not view the commercial promotion of the pacemaker with sympathy. They had not been offered shares in the business although they had contributed their ideas. Bücherl and his assistants responded by trying to develop an alternative pacemaker and it came to a breach between him and Schaldach.[22] The loss of this very fruitful and innovative cooperation with the necessary clinical user side therefore threatened continued innovation. However, by lucky coincidence a Biotronik pacemaker had been implanted in an East German patient during his visit to relatives in Würzburg, and Schaldach was able to establish a research cooperation with physicians at the East Berlin hospital Charité. Despite being located on the other side of the Iron Curtain, Schaldach developed several innovations with the East Berlin team. The most important product of this cooperation was the atrial active fixation lead, the continued development of aortic balloon pumps and a new biogalvanic energy source.

In October 1967, Franke and Schaldach decided to move their production facilities from Franke's flat to a house in the Neukölln district of Berlin. By the late-1960s it was clear that their original goal, to build a better pacemaker at lower cost had been met successfully. Production was well beyond ten pacemakers per day and they found that their original low prices (used as a way to win new accounts) could be increased so as to match the prices of competitors. Schaldach also turned to other distribution channels, the most important being Ad. Krauth, a large German sales organization that also made attempts to promote Biotronik products in neighboring countries such as in Scandinavia and the Netherlands.

The 1970s – a growing split between R&D and business operations

Biotronik had by the end of the 1960s developed a reputation for technically advanced and reliable pacemakers. In 1970, the same year as Schaldach was offered a professorship in applied physics at the University of Erlangen-Nuremberg, Biotronik had 22 employees and manufactured about 3,000 pacemakers per year, mainly for the German market. In order to keep the growth continuing, internationalization would be necessary. This fitted very well with Schaldach's newly found passion for flying: he could fly directly to international destinations, mainly in Western Europe. However, the most important market that he had his eye on was the booming one across the Atlantic.[23] Biotronik's initial entry into the US market was made in 1971 through an independent sales agent with sales activities led by an experienced former Medtronic sales representative, Steve Anderson. Due to Anderson's sales skills and Biotronik's smaller, lighter pacemakers in metallic encasing, the company's US sales rapidly grew to 3,000 units or 4.7 percent US market share on an annual base in mid-

1974.[24] Unfortunately, Biotronik experienced major problems with a cracking epoxy encapsulation of its pacemakers. A recall had to be issued, putting a stop to climbing market shares.

In the German market, Schaldach together with a doctor of medical engineering, Peter Osypka, in 1972 set up a semi-direct sales organization called Biomedix. Biomedix, which also had engineering and manufacturing capabilities proved to be able to handle the epoxy-caused recall extraordinarily well. This prompted Schaldach to set up a direct organization in the US to save customer confidence, look after Biotronik interests, and become independent of the distributor. Even if this initially was not a deliberate plan to expand complementary capabilities in sales, Schaldach quickly learned to appreciate the advantages in handling delicate end-customer relations and subsequently launched direct sales organizations in France, Austria, Greece, and England.

An additional challenge that hit Biotronik in 1974 was the advent of the lithium battery. Biotronik had made alternative choices, first a bio-galvanic energy source, then the Promethium radioactive isotope energy source. Schaldach had identified the energy source as the weak link in the cardiac pacemaker system very early but approached this as a purely scientific problem without reflecting enough upon the governing values of customers and regulating institutions. He therefore failed to perceive the potential of the lithium battery.[25] Newer companies quickly exploited this innovation in the US market, making it difficult for Biotronik to regain market share there. Expansion continued in Western Europe instead, where diffusion of the lithium-powered pacemakers was less pervasive.

Following the introduction of the lithium battery, the next big innovation was the programmable pacemaker. Biotronik was not able to take advantage of this innovation because the company lacked capability in advanced microelectronics design. Microelectronics represented a technological paradigm different from the discrete, analog electronics mastered by Franke and R&D director Dr Blaser. Blaser did not regard integrated circuits as reliable enough and Schaldach did not foresee demand for a non-invasively programmable pacemaker, even though the US firm Cordis had launched its programmable Omnicor pacer already in 1972. This device had been promoted as an alternative way to solve the problem of a short service life, as the physician could reduce the amplitude of the stimuli, thereby slowing the drain on the battery. The problem with a short service life, however, was soon solved by the lithium battery, therefore at once changing the meaning of programmability from a means of improving longevity to a way of flexibly optimizing pacemaker stimulation for each patient. This was an innovation appealing both to academic key opinion leaders and to regular implanters. Being able to quickly respond to the preferences of different customer groups had become absolutely essential in the rapidly changing US pacemaker market of the mid-1970s. As Biotronik lacked a marketing department this proved a clear weakness.

Schaldach, who saw his company's core strength in its technical capability to design small, reliable, and long-lasting pacemakers, had been beaten by the competition twice during the 1970s. His plan to strike with the Promethium battery had to be reversed by strengthening the company's focus on reliability. His position in the company as the only person who integrated several distributed activities had contributed to its inability to identify the increasing importance of specialized sales and marketing, and to its slow response to new emerging technologies. This lack of cross-functional and geographic integration, in addition to the weakness in marketing, reduced the innovative capability of Biotronik. By the mid-1970s Schaldach started a drastic renewal process by concentrating activities at a newly built headquarters in Berlin, replacing Franke as the managing director and Blaser as vice president of R&D, and buying a pacemaker development project in the US. Schaldach had received information that Johnson & Johnson (J&J) wanted to sell their CRM device arm. He set out to acquire J&J's programmable pacemaker project, including engineers with skills in digital logic design, custom-designed integrated circuits, and valuable patents. Personnel and resources were reorganized in 1979 as a subsidiary (Medical Technology Inc. or MTI) at a new site outside Portland, Oregon.

The 1980s – strengthening the technological roots of Biotronik

Biotronik entered the third decade of cardiac pacing with a fully overhauled organization and a leadership strengthened by Hermann Rexhausen as managing director of Biotronik and by Scott Shanks as general manager of Medical Technology Inc. in the US. Both would come to serve Schaldach and Biotronik loyally for their remaining professional lives and both were very able managers with a focus on getting things done. Rexhausen, with prior experience as a management consultant, brought stability to Biotronik by implementing organizational structures and processes with clearer personal accountability. Like Schaldach, Rexhausen was tough, entrepreneurial and achievement-oriented, personalizing old Prussian virtues of diligence, discipline and dedication.[26]

The first product of Biotronik after the reorganization was a single-chamber multi-programmable pacemaker, its electronic circuits developed by MTI in Oregon and its mechanical design and assembly done in Berlin. With this product Biotronik resumed its attempts at targeting the US market. Unfortunately, the extraordinary design measures taken to maximize its reliability and safety made the pacemaker bulky compared to the products of US competitors.[27] Just as reliability in the 1960s and longevity in the early-1970s had been the critical parameters for the perception of these devices by physicians, small size together with programmability were what signified an advanced pacemaker at the beginning of the 1980s. Especially in the US market it was almost impossible to sell a product that did not comply with such widespread expectations.

While Medtronic in the late-1970s and early-1980s shrewdly promoted pacemakers that stimulated both the atria and the ventricles (dual-chamber pacing) by coordinating clinical research activities with professional key-opinion leaders, regular users and regulatory agencies, Schaldach and Biotronik concentrated primarily on technical realization. Once the market trend was clear, the close and centralized leadership control exercised by Schaldach and Rexhausen allowed Biotronik to respond quickly. Within a year the company produced a new dual-chamber pacemaker, the Diplos 03, based on the chip set of the programmable single-chamber pacemaker, and simultaneously started a project to develop a new generation multi-programmable single-chamber pacemaker based on single-chip CMOS technology. The result of this project was Neos, the smallest single-chamber pacemaker at the time of its introduction in 1983. Neos was the result of three separate but cooperating development teams, each in a different organization and country: MTI in Oregon, Biotronik in Berlin, and SGS Thomson in Milan. Diplos, by contrast, had basically been the work of the MTI engineer Dennis Digby working closely with Schaldach and the new vice president for R&D in Berlin.

Technological innovation: key to success and failure

Both Diplos and Neos met with great commercial success in the early-1980s. Diplos, with its conservative specification and pioneering atrial-based timing cycles, turned out to be a very stable pacemaker without the problems of some competing models. The early-1980s saw many technical problems with pacemakers and leads, mainly due to difficulties in mastering new technologies that were introduced in order to keep up with competition. In this respect, Biotronik was very cautious, only introducing new products or process technologies after systematic evaluation either by the quality assurance department or research by doctoral students, which led to the decision, for example, not to use polyurethane as a lead insulation material.[28] This cautious policy largely explained why Biotronik was spared of product recalls in the demanding regulatory environment of the 1980s.

By the mid-1980s the pacemaker was redefined from a device for treating slow heart rhythm into an advanced system that adaptively monitored and influenced the heart rhythm. Third-generation dual-chamber pacemakers introduced in 1984 under the slogan of 'physiologic pacing' intensified competition by adding new features such as self-adjusting algorithms and advanced diagnostics. Pacemakers built around microprocessors for treatment of atrial tachyarrhythmias and the implantable defibrillator for treatment of ventricular fibrillation had been clinically introduced.[29] 'Physiologic pacing' included pacemakers that could change their rate according to the metabolic needs of the patient. Thin and slippery bipolar leads of new materials with low stimulation thresholds were also part of the concept.

The relentless pursuit of technological innovation that had started with the application of microelectronics and new types of cardiac stimulation required institutional reengineering both within firms and of the markets. The move to microprocessor-based devices was difficult for Biotronik as the technology challenged deep-seated beliefs: Schaldach, a physicist believing in universal natural laws as the scientific, stable base for technology, realized the potential of the microprocessor but was never comfortable with its complex and malleable, therefore fallible nature. The technology also shifted the focus of innovation from technical hardware matters to clinical software matters. Extensive marketing, including training, education, clinical studies and PR activities had become essential parts of innovation, but were shunned as costly and not subject to a scientific approach. Marketing at Biotronik was kept to a minimum in an environment that required extensive analysis of environmental trends, development of attractive concepts and clinical data to back them up, so as to synchronize Biotronik with the evolving dynamic industry innovation system (see Figure 7.1).

The emerging regime of continuous innovation was driven by market competition between companies in the central market sphere and academic competition between clinical research teams of the acad-

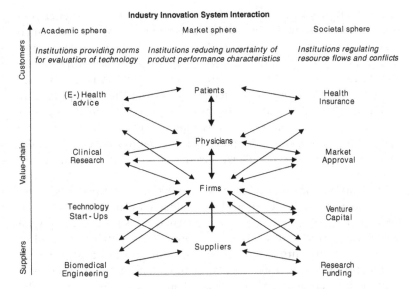

Figure 7.1 Graphical representation of interactions in the industry innovation system.

emic sphere. This 'horizontal' competition within each sphere at several levels forced cooperation both in a 'vertical' dimension between company representatives and University researchers and horizontally between actors across the spheres. The orchestrating actors of these research networks were the centrally positioned companies who coordinated activities between biomedical research, industrial engineering, clinical research as well as lobbied resource stakeholders and regulatory agencies. The smooth coordination of these networks required an ability to act in accordance with the dominant values and norms of various institutional environments, something that also required an internally structured and coordinated organization of many diverse capabilities.

Biotronik's R&D capacities in Berlin were clearly insufficient for the challenges presented in the mid-1980s. Several good ideas were only partly explored. Despite limited resources, in 1984 the company started a project for an improved electrode surface, initiated work on a new, fully automatic pacemaker and began building a plant in northern Bavaria for designing and producing hybrid circuits, the mechanical and electrical interconnections between the various discrete components and integrated circuits (ICs). In the same year Schaldach also assigned his new R&D manager the task to develop the next generation dual-chamber pacemaker as well as a rate-responsive pacemaker, which was to automatically vary its stimulus rate based on some sensed physiologic change in the patient. As this list suggests, many projects were started to keep up with competition, but project management procedures, resources and equipment for realizing these projects were scarce. In addition, conditions for projects involving several geographical sites were inadequate. As a consequence, projects did not develop successfully unless key participants made extraordinary efforts. In the next few years several people who were central to crucial projects left Biotronik.

The limited market success of Biotronik's temperature-based, rate-responsive pacemaker Thermos led Schaldach, in late-1987, to launch development of a rate-responsive pacemaker using a piezo-crystal sensor that translated mechanical energy as the patient engaged in physical activity into electrical signals controlling the pacemaker's stimulation rate. The huge success of the Medtronic Activitrax pacemaker, which enjoyed a global market share of nearly 20 percent at the time, was due to its simple design and fast response to patient physical activity and turned this approach into industry standard by the late-1980s.[30] The device both complied with the dominant norms of physiologic pacing supported by key opinion leaders and met the needs of general users who wanted a simple solution.

The products that would succeed the Neos- and Diplos pacemakers and would keep up sales momentum during the latter half of the 1980s were lacking. Despite this, Schaldach and Rexhausen continued to build additional internal technological capability in hybrid circuits at their US plant in Lake Oswego. Hybrid circuits were expensive to buy but critical for quality and for making pacemakers smaller. Biotronik was thus able to launch smaller versions of older pacemakers such as the 6mm-thin Mikros in 1986 and the Nanos in 1989 at lower cost and higher reliability. However, these lacked the advanced features of competitor products, making it difficult to gain new customers and market share except in low-price regions such as Eastern Europe and Latin America. Still, many older customers kept buying single-chamber pacemakers, which remained the largest product segment. Biotronik growth rates were lower in the second half of the 1980s, but substantial enough to expand production capacities by relocating the 228 employees to Woermannkehre 1 in Berlin in late-1987, where the firm still has its headquarters. This relocation was also used to rationalize production methods significantly. The inner assembly of feedthrough, case and battery were now interconnected by laser spot welding to enable a higher degree of process automation and miniaturization of the pacemaker.

1990s – diversifying Biotronik into related fields

In addition to competition in technology, device manufactures faced new layers of complexity relating to regulation and changing market conditions. In the US during the 1980s, companies were learning to live within the regulatory regime that had taken shape since 1976 under the auspices of the Food & Drug Administration (FDA, which assesses the safety and effectiveness of new devices) and the Centres for Medicare & Medicaid Services (CMS, the agency that administers Medicare and sets its reimbursement levels for medical devices and procedures).[31] The increasing regulatory requirements in the US market constituted a tremendous challenge to Biotronik in the 1980s as its focus was not on market development and commercialization of medical technologies, but on development, manufacturing and sales. The US regulatory requirements spread also to other countries in the 1990s. From 1998 device companies marketing products in Western Europe also had to meet the complex (and evolving) regulatory requirements of the European Union which supplanted the rules of individual countries.[32] Manufacturers planning to introduce a new implantable device had to assess whether the innovation would likely gain regulatory approval, whether it was compatible with standards of medical care, the training of physicians, the organization of hospitals and clinics in each country, and whether the innovative device would earn a profit.

In the 1990s, as hospitals and large clinics in the US consolidated and became more assertive in negotiations with suppliers, the device manufacturers feared downward spiralling prices for the first time. The growing purchasing power of health maintenance organizations (HMOs) and group purchasing organizations (GPOs) resulted in less focus on the clinical benefits of product features and pressure on prices that induced the US firms to adopt new strategies. A main strategy pursued by successful companies was diversification of their product-range, searching for acquisition targets as a way to add new technologies that promised high growth in the future, investing even more in R&D to maintain a stream of innovations and especially in clinical trials to demonstrate the clinical benefits of device features. The companies also aimed to 'offer providers one-stop shops through broadened product lines, and price concessions through increased volume.'[33]

Through acquisitions, Medtronic by the mid-1990s was able to offer a full line of cardiovascular products, not only pacemakers and defibrillators but an array of implantable leads, angioplasty catheters, stents, guide wires, heart valves, and related hardware, the tools of electrophysiologists and interventional cardiologists. The two other large US CRM companies that had formed through extensive acquisitions in the mid-1990s, Guidant and St Jude Medical, followed suit.[34]

Biotronik also considered a range of further diversification options including orthopaedic joints, heart valves, coronary angioplasty, ablation therapy and neurostimulators in addition to implantable defibrillators (ICDs). In contrast to its US competitors Schaldach contemplated organic diversification based on the know-how and contacts accumulated during his academic research in biomedical engineering. Due to restricted financial funds only ICDs, coronary angioplasty and ablation products would be commercialized on a larger scale. Although Schaldach saw the ICD as a crude technology, he did not overlook the fact that these devices were implanted in increasing numbers in patients who had survived a cardiac arrest. Most competitors had started designing their ICDs in the mid-1980s following antitachycardia pacemaker development. Biotronik, however, had never developed a microprocessor-based antitachycardia pacemaker and therefore lacked significant clinical and technological expertise that most competitors had been accumulating over a period of nearly ten years. Biotronik did not devote any research to defibrillators until 1989 yet was able to develop a defibrillator prototype within three years. This was even more impressive as most other competitors failed to introduce competitive defibrillators following the expiration of CPI's basic patent in 1990. The explanation for this achievement could be found in the development team of Dr Pilz, an East German who Rexhausen had been able to appoint following the fall of the Iron Curtain.

Two other projects initiated and driven by Rexhausen in the early-1990s were the development of electrophysiology (EP) and a PTCA (percutaneous transluminal coronary angioplasty) product lines. These were primarily catheters which the company produced with an extrusion machine purchased in the late-1980s. The EP business had a long tradition in the German sales organization where external pacemakers and diagnostic measurement equipment had been developed on a customer specific basis since the 1970s. Contacts made with an East German engineering company during the closed-loop pacemaker development in the 1980s were now utilized for the development of a therapeutic ablation generator. First experiments with the development of PTCA balloons were made in the early-1990s together with former East German researchers in Rostock. Despite initial problems with materials and patents, a small scale PTCA business got started 1995.

Schaldach and Rexhausen saw the diversification steps into ICDs, EP and PTCA as necessary steps to adapt to the increasingly demanding environment where cross-selling was increasingly important. Clinical affairs, marketing and sales developed into co-specialized complementary assets. Biotronik's strong technological absorption capacity and its excellent academic relationships made this diversification possible despite a lack of the strong financial base of its US competitors. The fall of the Iron Curtain in 1989 also constituted a tremendous opportunity to expand business and acquire medical technology know-how. Biotronik with its location in West Berlin had nurtured extensive contacts with physicians and researchers in East Germany in the 1960s and early-1970s. In 1990, Biotronik was able to acquire an East German lithium battery plant, thus further strengthening its technological independence.

Biotronik had earlier expanded its Latin American operations as competitors withdrew when national health systems in that region faced economic problems. In the early-1990s, Biotronik increasingly captured the East European and Latin American markets due to the company's readiness to continue delivering devices even if customers were unable to pay. Biotronik supplied older models of pacemakers at no financial profit, but customers in these regions were primarily interested in the basic pacing functionalities, not the latest features. These activities won the company new academic contacts and opportunities to test new products and concepts clinically. The ambitious research project on a pacemaker with closed-loop regulated stimulation, for example, underwent most of its clinical testing in Brazil. Research contacts in these countries were also crucial for the rapid clinical testing and development of Biotronik's Phylax implantable defibrillator, which was commercially introduced in 1995.

Device manufacturers in the 1990s, paradoxically, began to actively take advantage of regulatory requirements. Clinical research results from well-designed studies became increasingly important. The first example of this was the drastic expansion of dual-chamber pacemaker use following the publication of the first research results proving mortality benefits of physiologic dual-chamber pacing over single-chamber pacing.[35] With statistically significant and clinically relevant differences in 'hard' endpoints such as mortality, healthcare purchasers were pressed to pay for more expensive and advanced devices. Interestingly, this was especially the case in the highly institutionalized US CRM market. Following studies of dual-chamber pacemakers, the clinical advantage of ICDs compared to drug treatment was established in clinical studies financed by the CRM device industry. Toward the end of the 1990s device manufacturers turned their attention to congestive heart failure. They were building on their understanding of electrostimulation to introduce three-chamber pacemaker devices that resynchronized the pumping of the left and right sides of the heart, thereby improving the delivery of blood throughout the body. Hundreds of thousands of people are diagnosed with heart failure each year in the US and Europe; thus cardiac resynchronization therapy represented an important new growth opportunity for the CRM industry.

Biotronik in the mid-1990s: toward a more decentralized organization

The diversification efforts that began in the early-1990s made it obvious to Schaldach that the Biotronik organization needed to be put on a broader base and that this would require a new organizational culture of more widely shared responsibilities and a more cooperative management style. This became even clearer when Hermann Rexhausen, early in 1995 became seriously ill. The first occasion for Schaldach to present the new direction (called 'BIO 2000') was at a large company employee meeting in April 1995. In his speech during this meeting he argued for the need to introduce a more decentralized divisional organization.

Despite Rexhausen's management capabilities, it was no longer possible for two people to manage a worldwide company of more than 1,000 employees that was diversifying into several new product markets and at the same time had come under attack by a Medtronic patent infringement lawsuit. The significance of this patent litigation became clear to Biotronik in late-1994 when the first referee hearing the case determined that Biotronik had intruded on patents regarding threshold tests and marker channels. This came at a time when Biotronik was expanding its excellent position in emerging markets and product segments and its core business was steadily closing the gap with competitors. The company's rapidly developing product portfolio of pacemakers supported by an advanced programmer and excellent leads with fractally coated electrodes positioned Biotronik for further growth. Most importantly, the management of Biotronik had started to understand the importance of clinical studies as marketing tools and their decisive importance to gain entry to markets.

An important element of the BIO 2000 company directives was the intensification of clinical research cooperation with customers. Schaldach consequently expanded clinical research in Erlangen in addition to the newly created regulatory department in Berlin which organized clinical studies for product approval purposes. The clinical research groups in Erlangen mainly consisted of doctoral students with an interest in gaining industrial experience while working on their research projects on clinical uses of the new products. This was a clever way of recruiting new employees for the R&D-activities of Biotronik. These activities played a key role in clinical research on new innovative products such as the fractally coated electrode and a pacemaker that included a high-resolution analog telemetry chip to transmit intra-cardiac signals and identify heart transplant rejection. This pacemaker could in combination with the fractal electrode also be used for drug and arrhythmia monitoring, leading to several clinical studies of great scientific interest. The more concrete result of BIO 2000 was the intro-duction of three separate 'business units' for leads, implantable devices, and interventional cardiology. As in the 1970s, Schaldach had inter-vened in the hope of making the organization more capable of growth.

Biotronik at the end of the century: becoming a global player

The lawsuit over patent infringement filed by Medtronic against Biotronik at the Düsseldorf state court in 1993 was the start of an intensive dispute between the two companies that in 1997 resulted in a cross-license agreement. Biotronik's success in the courts was replicated in the worldwide markets. Biotronik's bradycardia pacing business profited from a wide range of interesting new products including a dual-chamber pacemaker with ventricular capture control, a pacemaker capable of stimulating both the atria and the ventricles using only a single-lead and the first generation closed loop stimulating pacemaker. The company's ICD with a battery produced at Litronik in Pirna was in 1997 complemented by an ICD with a larger memory for storage of arrhythmic events and a dual-chamber ICD. Biotronik's interventional cardiology business likewise grew rapidly based on a popular guide wire, the introduction of new PTCA catheters, and a new steel stent with an antithrombogenic amorphous silicon carbide coating.

The many interesting clinical concepts, which were explored, in several clinical trials in several countries contributed significantly to the success of Biotronik. However, the company also profited from the demise of two major competitors, Telectronics and Intermedics, which provided the surviving companies with opportunities to attract good salesmen and engineers. To fuel this success with ideas for the future and larger development capacities, Max Schaldach in late-1997 hired a team of Telectronics engineers to complete the Inos development. Work commenced on a telecardiology solution for pacemakers and ICDs during 1997, introduced to the market six years later in 2002. In parallel with these R&D developments, increased emphasis was placed on a more regulation-conform design process. Increased regulatory requirements were enforced by the FDA in the US and by the European Union through the medical device directives of 1994; both became effective during 1997–98.

The expanding business volumes and corresponding complexity due to diversification meant the organization had to develop accordingly. Rexhausen expanded financial control and an enterprise-wide resource planning system was introduced in steps. Simultaneously the manufacturing capacities for hybrid electronic circuits were expanded and the process automated in Lake Oswego, a project led by Schaldach's son (also named Max). Not only the technical capabilities and control systems needed to be expanded, team work was introduced in the production and 'management by objectives' was introduced at the middle-management level so as to give employees a clearer direction and larger autonomy. These were necessary steps taken to transform Biotronik from an essentially owner-employer led company to a global player, which was Schaldach's vision.

The importance of marketing in handling the regulative, normative and cognitive institutions that selected innovations as well as the need to nurture more entrepreneurs in the Biotronik organization had become clear to Schaldach in the latter half of the 1990s. Still, he and Rexhausen faced problems in building an organization that would be able to read, respond to and shape the expectations of customers and other stakeholders. There still was a long way to go before the necessary structures and processes were in place. However, the insight was there and progress had been made with the BIO 2000 policies.

A generational shift in top management

On 29 November 1998, Hermann Rexhausen, who for over 20 years had translated the visions of Schaldach into detailed projects, provided clear directions for employees in Berlin, maintained customer relations, and had been entrusted with solving difficult problems, died in an airplane accident. This was a tremendous loss to Biotronik, and Schaldach had to immediately install a management board with responsibility for finance, technology, sales and marketing. In parallel, Schaldach worked with his usual dedication to continue Biotronik's long-term growth and prepare the company for the future. By 1999, the interventional cardiology business of Biotronik was experiencing extraordinary growth and profitability. To enable this trend to continue, Schaldach bought a clean room manufacturing plant in Bülach, Switzerland, and built a production plant for coronary stents in Warnemünde outside Rostock. His most important project was to establish a foundation for interdisciplinary research in cardiovascular technologies. This was an investment for the future to ensure a steady stream of innovative ideas. Unfortunately Professor Schaldach was not able to finish this project: on 5 May 2001, tragically, he too died in a plane crash at the same airport in Nuremberg where Rexhausen's accident had occurred.

The deaths of Rexhausen and Schaldach meant the end to an era of founder-led entrepreneurship and exclusively top-down innovation and management. A tremendous challenge lay before the founder's son, Dr Max Schaldach, to continue the legacy of his father, but in his own manner. The first steps to create new management and innovation systems for Biotronik were taken immediately after he took up his position as owner-manager. Dr Schaldach brought in a team of management consultants to develop new initiatives. In cooperation with the top management of Biotronik, a program of projects aimed at building comprehensive company-wide structures and processes with a stronger market influence was initiated. This signalled an ambitious reinvention process to align Biotronik better to healthcare systems worldwide.

Discussion

40 years of German entrepreneurship in medical technology is a title that refers to both Max Schaldach and his company Biotronik. His entrepreneurship was rooted in an academic interest to explore and analyze the clinical utility of new technologies, often involving academic peers in his ventures, which often had both an academic and an industrial dimension. Schaldach did not accept outside pressures to choose between the two institutional worlds which he saw as intrinsically related. Research results that could not be realized were of minor importance to him. However, he did not play an integrative role that could nurture the accumulation of core capabilities. This role was

instead fulfilled by Rexhausen who often had to protect the operative core of engineering and manufacturing from too many uncoordinated impulses emanating from Schaldach who did not always adhere to the necessary institutional structures for industrial exploitation of an innovation to occur. What the Biotronik organization primarily lacked was a strong and integrated marketing function that could examine ideas by Schaldach and other innovators according to future business criteria based on data-driven analyses, and translate them into operative actions. Another weakness was the relatively limited financial resources of Biotronik compared to its US competitors which had become public companies earlier in their history. This limited the company's ability to make marketing investments that could have expanded operations and exploited innovations commercially in a rapidly changing market.

The resources of US companies were used to deliberately shape the expectations of customers and other stakeholders by a set of concerted marketing activities such as clinical studies, education, symposia, advanced promotional and communication strategies. Biotronik by contrast had problems complying strictly with requirements for clinical studies. This is one reason why Biotronik had difficulties competing in the US environment following the introduction of a set of regulative institutions in the 1970s and 1980s. Commercially successful innovation required not only technological capabilities but also the ability to understand the preferences of customers and the mechanisms behind the diffusion of innovations and penetration of markets. Schaldach often based his views on purely scientific data and not always on the criteria relevant to the needs of customers, taking into account professional opinion leaders, but at the expense of expectations forming among users. As he saw research and marketing as his main responsibility, alternative views rarely gained influence at Biotronik. The top-down mode of innovation and management practiced at Biotronik hampered cross-functional integration which is crucial in promoting innovation. Biotronik was good at technology, especially effective production, but lacked the crucial complementary assets to ensure a rapid acceptance of products in the market.

These weaknesses in the commercial exploitation of innovation, insufficient adherence to market norms, a lack of internal functional integration and insufficient marketing capabilities, largely explained why Biotronik was not successful in the US. This market required a strict adherence to institutions, foresight of and quick reaction to emerging opportunities and the capability to create new trends, a profound understanding of the requirements of the contexts of use and a strong, convincing market identity. On the other hand, Biotronik was fairly successful outside the US due to a set of other factors: first, the entrepreneurial ability to react quickly and flexibly to opportunities and mobilize the organization in response to external threats based on close top-down relations to middle management. Second, an excellent capacity to absorb new technological developments and technological capabilities due to close university ties. Third, close contacts to key opinion leaders in emerging markets through cooperation in clinical research. Fourth, a lean and cost-effective organization able to generate profits even in the lower-priced markets, and fifth, a decisive commitment to long-term success by an uncompromising willingness to invest in long-term research.

Much of Biotronik's success can be credited to Schaldach and Rexhausen. Despite the company's significant commercial achievements, Schaldach never quit his academic career in medically applied physics. This helps to explain the character of Biotronik as a founder-led company where both research and marketing were a function of Schaldach's personal enthusiasm for offering physicians new devices that would help them treat their patients better. He never gave up looking for technological approaches that would give him an opportunity to convince physicians of the superiority of his products over those of competitors. On the other hand, he never became intensely interested in the organizational and technical details of design, production and logistics; he delegated these activities to a managing director working closely with him. First Franke fulfilled this role for 15 years, then Rexhausen, who committed himself fully to the success of Biotronik. Schaldach's networks of university friends and associates, his entrepreneurial capabilities, and his ability to persuade his colleagues to work hard helped to make Biotronik more successful than many other European competitors. But Biotronik's history also shows that, as technological, organizational and institutional complexity increased over time, the relative importance of the individual decreased, and the importance of collective integrated action grew.

Acknowledgments

Many thanks to all the employees of Biotronik who shared their insights with me and to Professor Kirk Jeffrey at Carleton College, Minnesota for his many useful comments and extensive help in preparing this manuscript.

Notes

1　Medtronic Inc. is the biggest of the US cardiac rhythm management companies and is located in Minneapolis, Minnesota. The two other US cardiac rhythm management companies *Guidant* and *St Jude Medical* also have major operations in the twin cities of Minneapolis and St Paul.

2　D.E. Zinner, 'Medical R&D at the Turn of the Millennium,' *Health Affairs*, 20 (2001), 202–9. Corporate data of the US Securities & Exchange Commission indicate that Medtronic, *Guidant*, and *St Jude Medical* collectively invest about 11 percent of net sales in R&D. Information for Biotronik a privately held company, is not available.

3　S.B. Foote, *Managing the Medical Arms Race: Innovation and public policy in the medical device industry* (Berkeley, University of California Press, 1992); K. Jeffrey, *Machines in Our Hearts: The cardiac pacemaker, the implantable defibrillator, and American health care* (Baltimore: Johns Hopkins University Press, 2001).

4　M. Moran, *Governing the Health Care State: A comparative study of the United Kingdom, the United States and Germany* (Manchester: Manchester University Press, 1999), pp. 110–19.

5　P.J. DiMaggio and W.W. Powell, 'The Iron Cage Revisited: Institutional isomorphism and collective rationality in organizational fields,' *American Sociological Review*, 48 (1983), 147–60.

6　J.M. Utterback, *Mastering the Dynamics of Innovation* (Boston: Harvard Business School Press, 1994); M.L. Tushman and C.A. O'Reilly III, *Winning through Innovation: A practical guide to leading organizational change and renewal* (Boston: Harvard Business School Press, 2002).

7　W.R. Scott, *Institutions and Organizations* (Second edition, Thousand Oaks: SAGE Publications, 2001), p. 52; P. Hidefjäll, *The Pace of Innovation: Patterns of innovation in the cardiac pacemaker industry* (PhD Thesis, Linköping: Linköping University, 1997), pp. 49–57.

8　M.L. Tushman and L. Rosenkopf, 'On the Organizational Determinants of Technological Change: Towards a sociology of technological evolution' in: B. Staw and L. Cummings (eds), *Research in Organizational Behavior*, 14 (Greenwich, CT: JAI Press, 1992), 311–47.

9　Hidefjäll, *The pace of innovation*, pp. 280–91.

10　The famous economist William Baumol speaks of the innovation machine: W.J. Baumol, *The Free-Market Innovation Machine: Analyzing the growth miracle of capitalism* (Princeton: Princeton University Press, 2002).

11　Yet observers look upon further progress with some trepidation because implanting and managing these halfway technologies (which control symptoms but can never cure the underlying disease) is believed to add to the cost of health care; see L. Baker, H. Birnbaum, J. Geppert, D. Mishol, and E. Moyneur, 'The Relationship between Technology Availability and Health Care Spending,' *Health Affairs Web Exclusive* (5 November 2003), available at http://content.healthaffairs.org/webexclusives/.

12　To cite one example, the US firm American Optical (AO) introduced a pacemaker able to track intrinsic electrical activity in the heart and inhibit itself from firing an artificial stimulus when it sensed a normally conducted ventricular beat, so that it paced the heart only 'on demand.' Awarded a broad patent in 1967 and introduced in 1968, the American Optical 'demand' pacemaker forced all other companies to develop similar devices, but AO itself never gained a sizeable market share in pacemakers and exited the cardiac rhythm management devices industry in 1975. See: Jeffrey, *Machines in our hearts*, pp. 155–9, p. 168.

13 M. Eden, 'The Engineering-Industrial Accord: Inventing the technology of health care,' in: S.J. Reiser and M. Anbar (eds), *The Machine at the Bedside: Strategies for using technology in patient care* (New York, 1984), pp. 49–64; T.P. Hughes, 'The Evolution of Large Technological Systems,' in: W. Bijker, T.P. Hughes, and T. Pinch (eds), *The Social Construction of Technological Systems* (Cambridge Mass., MIT Press, 1987), 51–82, pp. 64–6.

14 R.R. Nelson, 'The Co-evolution of Technology, Industrial Structure, and Supporting Institutions,' *Industrial and Corporate Change*, 3 (1994), 47–63, p. 54.

15 D. Teece and G. Pisano. 'The Dynamic Capabilities of Firms: An introduction,' *Industrial & Corporate Change*, 3 (1994), 537–56.

16 D. Teece, 'Profiting from Technological Innovation: Implications for integration, collaboration, licensing and public policy,' *Research Policy*, 15 (Elsevier, 1986), 285–305, p. 288; see also M.L. Tushman and P. Anderson, 'Technological Discontinuities and Organizational Environments,' *Administrative Science Quarterly*, 31 (1986), 439–65.

17 Nelson, 'Co-evolution,' pp. 55–7.

18 E.S. Bücherl and D. Lidgas 'Die elektrische Reizung des Herzens mit sogenannten Schrittmachern – Technik, Klinik und Problematik,' *Berliner Medizin*, 15 (1964), 481–9.

19 Discussion with Professor Max Schaldach and interview with Professor Ha[o]kan Elmqvist whose father led the R&D lab at Elema-Schönander at that time.

20 A brief history of Biotronik and a biography of Max Schaldach (1936–2001) written by the author of this paper may be found at the Biotronik website, http://www.BIOTRONIK.de/. Some of the following analysis of Biotronik is based on discussions that the author had with Professor Schaldach prior to his death in an airplane accident in May 2001 and interviews carried out during 2001 as part of a Biotronik internal history project led by the author.

21 The three-letter pacing code classifies different types of pacemakers according to where in the heart they stimulate (V = ventricular, A = atrium, D = dual) and how they respond to a sensed electrical heart event (T = trigger, I = inhibit, D = dual). A VAT-pacemaker stimulates the ventricle, senses electrical activity in the atrium and triggers a stimulus in the ventricle after a sensed event in the atrium.

22 Interview with Ms Ingrid Wittenburg, medical-technical assistant of Dr Bücherl, 6 March 2003; H-J Wanjura, Konstruktion und Implantation eines aufladbaren Herzschrittmachers, *Zeitschrift f. Kreislaufforschung*, 58 (1969), 23–5.

23 The US market grew over 30 percent annually in unit sales and even more in dollar sales.

24 The US unit sales were 51,000 in 1972 and 64,000 in 1974 according to D. Gobeli, *The Management of Innovation in the Pacing Industry* (PhD thesis, University of Minnesota, 1982), Appendix A.

25 Jeffrey, *Machines in Our Hearts*, 169–74.

26   Colleagues who worked with Rexhausen often referred to him as 'tough' with a Prussian character.

27   N.P.D. Smyth, D. Sager and J.M. Keshishian, 'A Programmable Pulse Generator with High Output Option,' *Pacing and Clinical Electrophysiology*, 4 (1981), 566–70.

28   E. Weimer, 'Untersuchung der Stabilität von Implantatwerkstoffen mit Hilfe der Raman-Mikroanalyse' (PhD thesis, Friedrich-Alexander-Universität Erlangen-Nürnberg, 1984).

29   A tachyarrhythmia is a disordered heart rhythm in which the heart beats too fast while a bradyarrhythmia involves an unduly slow heartbeat. The normal range for the heartbeat in an adult who is awake but not exercising vigorously is defined as 60–100 beats per minute.

30   A.A. Lappen, 'Took a licking but keeps on ticking,' *Forbes* (3 October 1988), 176–7.

31   For an introduction to the way CMS determines whether to initiate coverage for a new medical device and determines a payment level, see *Health Care Industry Market Update: Medical Devices & Supplies* (Centers for Medicare & Medicaid Services, 10 October 2002), pp. 14–19. This report is available at http://www.cms.hhs.gov/.

32   For devices to be sold in the EU, manufacturers look to the Active Implantable Medical Device Directive (93/42 EEC) and its many subsequent modifications.

33   C. Littell, 'Redefining Product Value in Today's Market,' *Medical Device & Diagnostic Industry Magazine*, 18 (July 1996), p. 46. For an overview of the consolidation of the medical device industry with particular attention to Medtronic, see D.J. Morrow, 'Consolidation among Medical-Device Makers,' *New York Times*, 22 September 1998.

34   Today the product portfolios of both *Guidant* and *St Jude Medial* cover all CRM products but also other cardiology device products. Their respective product portfolios can be found on their respective home pages: http://www.sjm.com and http://www.guidant.com.

35   One of the first studies on mortality effects of pacing modes in patients was published by Swedish physicians comparing AAI and VVI pacing: M. Rosenqvist, J. Brandt and H. Schüller, 'Atrial Versus Ventricular Pacing in Sinus Node Disease: Effects of stimulation mode on cardiovascular morbidity and mortality,' *Am. Heart J.*, 116 (1988), 16–22.

8
Building Science-based Medicine at Stanford: Henry Kaplan and the Medical Linear Accelerator, 1948–1975

Takahiro Ueyama and Christophe Lécuyer

Introduction

The hospital is one of the places where people encounter highly advanced science and technology in their daily lives. In the hospital, patients are in contact with a number of scientific innovations and new machines and technologies, more so as computed axial tomography scanners (CAT) and magnetic resonance imaging scanners (MRI) have become indispensable tools for medical diagnosis.[1] The instrumentalization of diagnosis and hospital treatment, however, has not been without mixed results. First, the introduction of machines on the bedside, it is often alleged, has distanced physicians from their patients. Patients have found that machine-dominated treatment is given a privileged position and have felt alienated from the doctors' personal care. On the physicians' side, their clinical experience has been superseded by information derived from medical instruments. Furthermore, technologies in the hospital have often been associated with the ever-increasing cost of medical care.[2]

As many historians have argued, the United States represents one of the earliest and most dramatic examples of this instrumentalization of medical care.[3] After World War II, in particular from the 1950s onward, American medicine began to rely heavily on pre-clinical examinations carried out by way of laboratory technologies requiring the collaboration of other scientific fields such as physics, biology and chemistry. Biophysics, biochemistry, and genetics began to be seen as having an important contribution to make to medical care. The collaboration between medicine, physics, and engineering also became apparent:

since the 1950s many kinds of medical instruments have been invented through the collaboration of doctors and engineers and have been introduced in university hospitals.

Instrumentation is also related to the burgeoning collaboration between medical research and industry. As Stuart Blume argues, the intertwining interests of medical practitioners and large segments of the medical device industry fostered the process of technological innovation in medical instruments, especially in diagnostic imaging.[4] Nathan Rosenberg has also shown that key medical devices such as the endoscope were first developed by collaborative teams of physicians, physicists and engineers working in the same university and that these new innovations were often commercialized by start-up firms.[5] In addition, the entry of industry as an actor in this instrumentation process, by creating a standardizing platform, was supposed to create a new domain in medical practice.[6]

But to understand the emergence and efflorescence of science-based medicine in America, we need to re-examine the ways in which new instruments were used and constructed for medical practice. How are scientific machines brought into the hospital and transformed into diagnostic and medical tools? Who initiates the use of these technologies? What is the respective role of doctors, physicists, and scientific instrument companies in the introduction of new medical technologies? More generally, how did science-based machines transform medical practice and the post-war hospital in the US?

To explore these questions, this chapter focuses on the case of the medical linear accelerator (henceforth MLA), and more specifically on the activities of Henry Kaplan, a physician, biologist, and professor of radiology at Stanford University. Henry Kaplan is best known as a co-developer of microwave linear accelerators for cancer therapy (product name: Clinac). The Clinac was developed jointly by Kaplan, then Head of the Department of Radiology at Stanford University Hospital; Dr Edward Ginzton, a physics professor at Stanford; and Varian Associates, a Silicon Valley electronics component and scientific instrument manufacturer. Kaplan transformed this machine into a highly effective clinical tool for the treatment of Hodgkin's disease and other forms of cancer.

In this chapter we will show that Kaplan's bio-radiological research, together with his invention of MLA, was intimately intertwined with his enthusiasm for constructing science-based medicine at Stanford. In the post-war period, Kaplan developed a new approach to radiotherapy that combined clinical practice with experimental research and the development of new medical instruments. Using the new MLA, Kaplan and his group in the radiology department developed revolutionary protocols for the treatment of Hodgkin's disease. Perhaps more importantly, Kaplan played a significant role in the transformation of medical care at the national level – especially in radiation therapy. He helped build radiotherapy into a thriving medical specialty and a critical discipline for the treatment of many cancers.[7]

Kaplan's experimental research and clinical practice

Henry Kaplan was one of the pioneers of the new science-based radio-oncology in the US. Kaplan was trained in medicine at Rush Medical College in Chicago in the late-1930s, and did his residency in diagnostic and therapeutic radiology at the Michael Reese Hospital in Chicago. He also obtained a master's degree in radiology at the University of Minnesota in the early-1940s. In 1944, he joined Yale's radiology department as an instructor before working on mouse leukemia at the NIH. In 1948, he was offered the chairmanship of the radiology department at Stanford's medical school in San Francisco.

Kaplan's earliest research was devoted to the question of whether exposure of mice to total body X-irradiation (TBI) would induce leukemia. As early as 1945, in his first article in *Science,* Kaplan showed that the exposure of mice to TBI caused lymphomas; more interestingly, in his subsequent articles, he demonstrated that those X-rays induced leukemia in the thymus gland, which then spread to the lungs, spleen, liver, kidneys, and lymph nodes. One of his major findings was that the removal of the thymuses dramatically reduced the appearance of leukemia in these mice upon exposure to TBI.[8]

These animal experiments led Kaplan to investigate potential cancer viruses in leukemia. In an experiment of the 1950s Kaplan exposed thymus-removed mice to high doses of TBI, then grafted non-irradiated thymuses from other mice onto them. The result was that thymic lymphomas developed in the normal, non-irradiated thymuses of these irradiated mice. Finding that leukemia developed even in the non-irradiated grafted thymic cells suggested to Kaplan that leukemias and lymphomas in mice were induced by a latent leukemogenic virus triggered into action by physical and chemical agents. Thus, he argued, lymphoma on the thymus occurred not because of mutation but as a result of the activation of a potent leukemogenic virus.[9]

What distinguished Kaplan from other contemporary radiologists, however, was his combination of basic research and clinical practice.

'I felt,' Kaplan later recalled, 'that it was very important for diagnostic and therapeutic radiology to have an experimental as well as clinical base. I insisted on having a laboratory from the start.'[10] Kaplan also strongly encouraged his faculty to perform basic as well as clinical research. For example, in collaboration with Kaplan, Herbert Abrams, a diagnostic radiologist, developed new angiocardiographic techniques to investigate congenital heart disease. Faculty members who were not actively involved in research were weeded out. This emphasis on research was rare in radiology in the United States at that time. Kaplan's operation was one of the rare radiology departments with a sizable research program in the 1950s.[11]

To fund research, Kaplan turned toward the National Institutes of Health (NIH). The 1950s saw rapid growth in the extramural program of the NIH. Under the prodding of the medical lobby and with congressional support, the NIH budget for external grants grew from a few thousand dollars in 1948 to more than $100 million in 1957. Kaplan, who joined the peer review committees of the National Cancer

Figure 8.1 Sponsored research in the radiology department at Stanford.

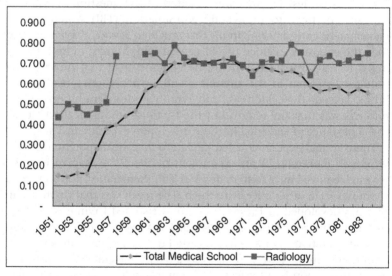

Note: Percentage of NIH funds in the overall operational budget of the radiology department and the medical school.
Source: Compiled from the Stanford Financial Reports and data in SC100 Encina Financial Documents, Stanford Special Collection, Administration, Box 5, Folder 5–19.

Institute (a branch of the NIH) in the early-1950s, had first-hand knowledge of the Institute's priorities. He developed an uncanny knack of finding NIH programs that would benefit his department. Under his leadership, the sponsored research budget of the radiology department increased from $32,080 in 1951 to $902,880, 15 years later. As a result, the department became increasingly dependent on outside research funds. The share of outside funding in the department's overall operational budget grew from 42 percent in 1951 to 75 percent in 1957 and 80 percent in 1964.[12] One of the main beneficiaries of this influx of research monies was Kaplan himself. In the 1950s, with NIH funds, Kaplan assembled his own research group in radiobiology and oncogene research.[13]

In conjunction with this experimental program, Kaplan developed an active clinical practice. His radiobiological experiments on mouse leukemia led him to do clinical work on Hodgkin's disease. Hodgkin's disease is a rare cancer of the lymph nodes. This cancer was almost always fatal at the time. To cure patients with Hodgkin's disease, in 1950 he pioneered a new treatment protocol. One of the main characteristics of this protocol was the use of very high doses of radiation. This was an innovative move. In the US at the time, radiation was used in small doses as a palliative technique to relieve symptoms rather than as a way to cure patients. Kaplan believed in the curative potential of radiation and administered high doses to his Hodgkin's patients (around 6,000 rads). This was five to ten times more radiation than was used in standard treatments.[14]

Kaplan's vision of combining therapeutics with scientific research was linked to his energetic involvement with reshaping the Stanford School of Medicine and moving it from San Francisco to its present location on the Palo Alto campus. Stanford's medical school at that time was no more than a regional school, which was dominated by clinicians and had a lackluster reputation. Rare were the faculty members in the clinical departments who were involved in basic research. Years later Kaplan recalled, 'the old guard in the medical school really were a bunch of pure clinicians, all very good in their fields, but with essentially zero track record in research or anything that could be remotely considered research.'[15] Henry Kaplan helped revolutionize this traditional clinically oriented culture and transform medicine at Stanford into a research-oriented, interdisciplinary field inextricably connected to the biological and physical sciences. Together with Wallace Sterling, then president of the University, and Frederick Terman, then provost, Kaplan was instrumental in Stanford's decision in 1953 to move the Medical School and its Hospital from San Francisco to the Palo Alto campus.[16] Against senior faculty members like Arthur Bloomfield, Harold Faber, and Emile Holman who believed that the relocation would curtail bedside clinical practice, Kaplan and other youngsters strongly supported the move. Citing ever-increasing dependence of medical research on the biological and physical sciences and engineering, Kaplan believed the move would promote collaboration between medicine and other sciences. A good example of this collaboration was his own work on the medical linear accelerator.

The MLA and the collaboration of medicine and physics

Kaplan's work on Hodgkin's disease made him acutely aware of the limitations of X-ray equipment then used in radiation therapy. Most machines produced low voltage X-rays which could not reach deep-seated tumors. These machines also had the disadvantage of concentrating much of the radiation on the patient's skin. This led to severe burns and prevented their use for long periods of time. Supervoltage X-rays would solve these problems. But, as Kaplan realized, the new supervoltage machines such as the Van de Graaff generator and the betatron, which were introduced into radiotherapy at that time, had clear limitations.[17] The Van de Graaff generator, developed by Robert Van de Graaff and John Trump, two MIT professors, and produced by their start-up company, High Voltage Engineering, produced high voltage X-rays through electrostatic means. The machine was difficult to use and highly unreliable. The betatron was also bulky and cumbersome. It required high voltages and was very expensive to install and operate.[18]

In 1951 Kaplan, searching for a new high-energy device for clinical radiation therapy, learned that Edward Ginzton, a physics professor and director of the Microwave Laboratory was creating a large-scale linear accelerator for high energy physics. In the fall of 1951, Kaplan wasted no time in setting up a lunch meeting with Frederick Terman, Edward Ginzton, and Leonard Schiff, chairman of the Stanford Physics Department. Kaplan later recalled:

> One of the nicest things about [the linear accelerator built by the microwave lab] was that you could get very high energies out of it for a very low energy input. You accelerated electrons with these high energies and you could hit a heavy metal target and make them into X-rays. I explained to Ed that's what radiotherapists were

for years dreaming about. We needed higher energies than we could possibly get from the crude devices then available. [..] This sounded miraculous and I became convinced that this was to become the radiotherapy machine of the future. Not only had I convinced myself, but by the end of that luncheon I had convinced all of them.[19]

Kaplan quickly proposed building a smaller, modified version of Ginzton's accelerator for medical use in the hospital. Kaplan considered that a million-voltage X-ray beam from the accelerator would penetrate to and kill certain deeper tumors difficult or impossible to remove by surgery or conventional X-rays. High-radiation therapy up to that time was more advanced in the UK. Kaplan himself recollected that his invention of the medical accelerator was 'six months behind the British'.[20] Ginzton visited British hospitals in 1956 to see the application of medical accelerators in the wards. He was impressed by the wide application of this technology; however he found that most of the radiation operations were carried out by Cobalt 60, Van de Graaff or particularly Betatron machines, all of which were able to produce only 2 to 3 million volts and – more of a liability – were very big, clumsy and expensive to use in the ward.[21] The linear accelerator, Kaplan believed, would be smaller and cheaper and have higher electron output, which in turn would make treatment shorter and thus more comfortable for the patients.

Ginzton and the physicists at the Microwave Laboratory enthusiastically embraced Kaplan's proposition. The new project would enable them to expand the scope of the laboratory and work on technologies that had a beneficial use. For almost two years, Ginzton and Kaplan struggled to raise the funds to design and build the medical accelerator. They applied for a grant from the newly established National Science Foundation. But Trump, who wanted to protect his business with Van de Graaff generators, used his position as a referee for the NSF to persistently reject their proposal. Ginzton and Kaplan finally received a grant of $68,000 from the US Public Health Service. This was in mid-1952. By 1956, they had obtained $100,000 from the American Cancer Society and $113,000 from the NIH. The James Irvine Foundation also gave $75,000 to install the completed machine in the Stanford hospital in San Francisco in late-1955.[22]

It is important to note here that building this medical instrument and applying it to medicine was mostly initiated not by doctors but by physicists and engineers. According to Ginzton, 'Kaplan did not care

Figure 8.2 Photograph of the Stanford Medical Linear Accelerator and the first patient, Gordon Isaacs.
Source: Special Collections, Lane Medical Library, Stanford University.

what voltage we would use. [The decision to use 6 MeV] was not on the basis of medical logic but just on engineering experience.' Kaplan told Ginzton, 'I don't care what you do. Anything between 2 and 10 million volts would be OK.'[23] Ginzton thought that with the existing linear accelerator technologies making a machine that produced very powerful X-rays would not be difficult but that the problem was in transforming 'what had been a large clumsy stationary machine into a compact device, easily movable in space, so that the doctor could position it quickly, easily, accurately and with knowledge that distribution of dosage within the body would be exactly as desired.'[24] In fact as we indicate below, the simple idea of making the machine compact, though it sounded easy, required a number of practical innovations.

The first MLA was built and tested in Ginzton's Microwave Laboratory and installed in the Stanford University Hospital in 1957.[25] Here, physicists and engineers played a more critical role than generally anticipated. It is interesting to note that when the first patient, a seven-month old boy, Gordon Isaacs, was treated for his retinoblastoma in 1956, the medical procedure of calibrating and focusing the radiation beam to the tumor was carried out not by doctors but by a team of physicists. The problem, it seems, was that 'When the doctors gathered around the machine just before treatment was to begin, they realized they did not know how to pinpoint the X-ray beam onto the tumor without destroying the rest of the retina.' On the spot, Ginzton devised the solution. By scotch taping a photograph of the eye to the child's temple, the doctors had a 'map' with which to aim the beam precisely, and the treatment was completed successfully.'[26]

The MLA was composed of five elements: (1) a cathode gun, the function of which was to produce a pulsed beam of electrons, (2) a radio-frequency device that amplified at high frequencies, (3) the accelerator wave-guide, a copper tube containing many loading disks that controlled the propagation velocity of high-frequency electromagnetic waves, (4) a gold or tungsten target that transformed a stream of electrons into X-rays, and (5) a mechanical mount to X-irradiate the patient's affected parts. The electrons discharged from the cathode gun would travel in an intense electric field propagated by the klystron in the wave-guide. If the dimensions of the wave-guide, the spacing of the disks, and the size of the holes were properly adjusted, an electron would travel in a sine curve on the phase velocity of electric waves as if it was surfing on an ocean wave, and would be accelerated to the velocity of light with the 2 MeV accelerator. When this accelerated electron hit a target like gold or tungsten, an electron was ejected from the outer electron orbit of the atom to the inner one and discharged as X-ray.

The first innovation of the medical accelerator was that the accelerator's wave-guide was sealed off. To avoid any disturbance in the electrons' path, the wave-guide had to be kept completely evacuated. For that purpose, powerful vacuum pumps had been attached to the physics research linear accelerators, but these pumps made it difficult to downsize the linear accelerator for medical applications. Similarly, the disk-loaded structure of the wave-guide made it extremely difficult to keep the joints between the wall and disks completely welded. The accelerator disks needed to be produced from oxygen-free copper, whose baking process with the inside wall was made very unstable under the heavy stress of a temperature much higher than the annealing point. Ginzton's team succeeded in solving these difficulties by using a new electroplating technology called 'electroforming' – his second innovation. The new method used aluminum spacers which were assembled alternately on a mandrel, then immersed in a copper-plating tank. After electroplating, the assembly was placed in a chemical tank, where the aluminum spacers were dissolved away. In so doing, Ginzton was successful in creating a completely sealed off wave-guide. Another innovation Ginzton was able to use was a super-high-frequency generator, the klystron that Hansen and the Varian brothers had invented for radar applications in the late-1930s and further developed in the late-1940s and early-1950s. In contrast, British manufacturers had no other choice than using an old-fashioned magnetron.[27]

These engineering innovations, to be sure, broke new ground in advancing diagnosis and therapeutics, but in the 1950s Kaplan's clinical protocol with a new machine faced very strong resistance from within the medical community. This exemplarily signifies that American medicine was in a time of transition: It was far from thoroughly accepting the specialization and 'technologization' of medicine. As Kaplan recalled later, radiologists of the day were afraid of irradiating large areas above and below the diaphragm with high-dose X-rays. As a matter of fact, many local radiologists, fearing the centralization of medicine in university hospitals, had severe qualms about Kaplan's therapeutic use of the linear accelerator. Ginzton recalled:

> In 1950, most cancer therapy was carried out in the private offices of doctors, and if a big machine were to be built, which could not be accommodated due to its cost and its size in the doctor's office, it would undermine private practice. The American Medical Association was very concerned about the transfer of the dramatics of treating cancer from private practices to institutional practice. There were all kind of articles written in San Francisco papers and [publications of] the Medical Association which sought to undermine Kaplan's position and the particular approach that he took.[28]

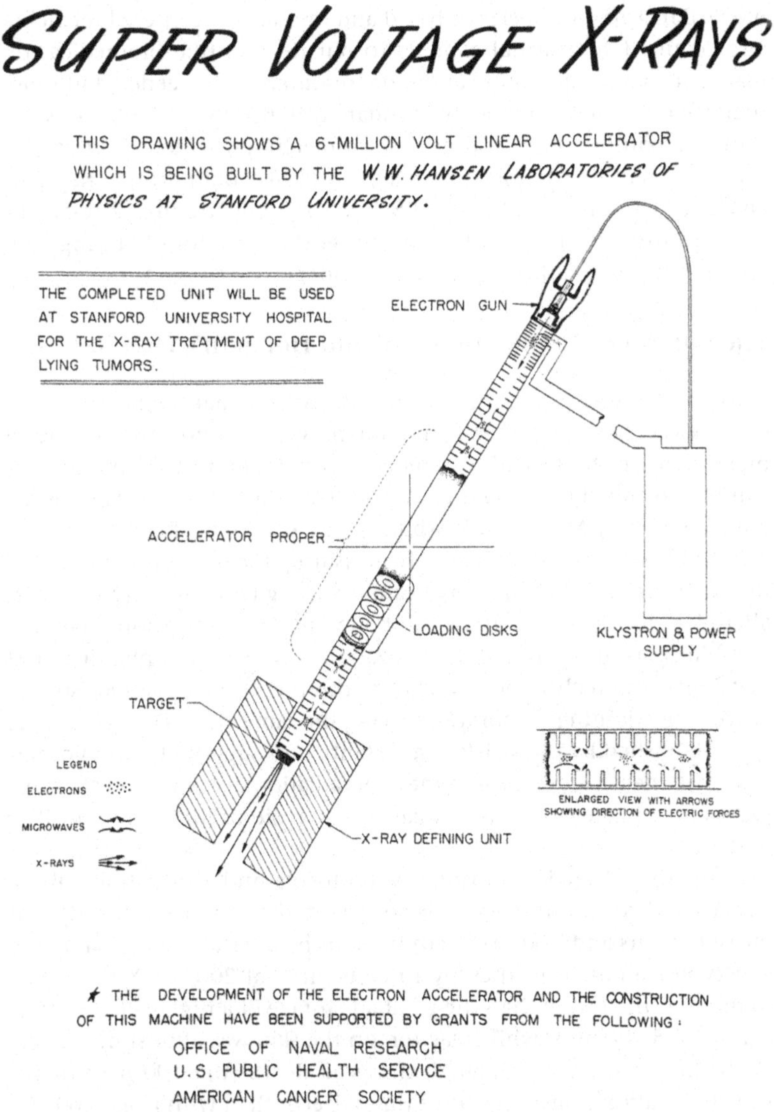

SUPER VOLTAGE X-RAYS

THIS DRAWING SHOWS A 6-MILLION VOLT LINEAR ACCELERATOR WHICH IS BEING BUILT BY THE **W. W. HANSEN LABORATORIES OF PHYSICS AT STANFORD UNIVERSITY.**

THE COMPLETED UNIT WILL BE USED AT STANFORD UNIVERSITY HOSPITAL FOR THE X-RAY TREATMENT OF DEEP LYING TUMORS.

ELECTRON GUN

ACCELERATOR PROPER

LOADING DISKS

KLYSTRON & POWER SUPPLY

TARGET

LEGEND
ELECTRONS
MICROWAVES
X-RAYS

X-RAY DEFINING UNIT

ENLARGED VIEW WITH ARROWS
SHOWING DIRECTION OF ELECTRIC FORCES

✦ THE DEVELOPMENT OF THE ELECTRON ACCELERATOR AND THE CONSTRUCTION OF THIS MACHINE HAVE BEEN SUPPORTED BY GRANTS FROM THE FOLLOWING :
OFFICE OF NAVAL RESEARCH
U. S. PUBLIC HEALTH SERVICE
AMERICAN CANCER SOCIETY

Figure 8.3   Schematic drawing of the Medical Linear Accelerator.
Source: Edward Ginzton Papers, Special Collections, Stanford University.

But as the development of new medical technologies by Kaplan and Ginzton exemplifies, American medicine in the 1950s was in the process of moving towards such centralization – for the university hospital made it possible to integrate research and therapeutics through medical instruments. Science-based and instrument-centered medicine also required substantial capital. To pursue scientific biomedical research at universities, more and more resources were needed, but more than that, the new therapeutic standard initiated by scientific research projects required large and expensive machines, which only capital intensive centers such as university hospitals could afford. Kaplan's endeavor to transform Stanford School of Medicine into a research-oriented institution, together with his enthusiasm for obtaining government funding, mirrored these developments.

The radiation therapy protocol and Hodgkin's Disease

Until 1961 Kaplan's treatments with the MLA remained both preliminary and experimental. At first, Kaplan's group performed dosimetry measurements to provide therapists with accurate information regarding absorbed doses. In the mid-1950s, four physicists who joined Kaplan's group, Michael Weissbluth, C.J. Karzmark, R.E. Steele, and A.H. Selby, succeeded in accurately measuring the distribution of radiation with the accelerator X-ray beam. By using two new devices, a parallel-plate ionization chamber and a cylindrical ionization chamber, Weisbluth's team measured the ionization within a water phantom and developed the techniques to acquire the precisely measured isodose curve, a line joining all points receiving the same dose. This technique, first established by physicists, was essential for supervoltage radiotherapy since the accurate measurement of radiation distribution was indispensable to make irradiation equal and to avoid irradiation to healthy organs.[29]

Using the physicists' dosimetry protocols and calibration tables, Kaplan and his group were able to devote the year 1956 to physical measurements and isodose determinations by clinical trials. Comparing supervoltage radiation therapy to conventional 200 KV X-rays, they found that the former had the great advantage of a higher-depth dose, 5cm with 5,900 rads, while reaction on the skin was almost negligible. With the MLA, they estimated, the peak depth (the 100 percent absorbed radiation point) was three times deeper than with Cobalt 60.[30]

With the prodigious machine in hand, Kaplan's new treatment protocols became very bold and experimental. In 1956, Kaplan and his group treated 74 cases with the 4 MeV MLA and operated the machine on 23 different kinds of tumors including those of the cervix, ovary, prostate, kidney, and tongue. More importantly, the dosages delivered to patients were very high, in amounts as great as 6,000 to 7,000 rads over six to seven weeks. Considering that Kaplan concluded in a 1966 article that the most suitable doses for radiation therapy were between 4,000 and 4,500 rads, his treatments at least until 1961 were very experimental in nature.[31]

In 1961, with NIH grants of $943,412 for the first year and $600,000 per year for the next 6 years, Kaplan and his group established the Clinical Radiotherapy Cancer Research Center at the Medical School and initiated fully fledged cancer treatment with the MLA.[32] From then on, their clinical trials began to focus on particular tumors in the throat, uterus, prostate, and bladder, which were particularly deep and could not be properly treated by traditional machines. Then in 1962 Kaplan moved to the treatment of his main target, Hodgkin's disease.

Here again the influence of physicists and engineers was very important in that their technological insights seem to have helped Kaplan focus on particular tumors. As we have seen, the original model of the MLA was particularly suited for treating cancers of deep tumors in the head and neck, prostate, and bladder. As Greg Nunan, a physicist at Varian Associates later recalled, the advantage of this machine lay in its ability not only to produce X-ray beams but also to emit electrons directly. Unlike ordinary X-rays, the electrons could penetrate the body to a certain distance and stop at the tumors without irradiating other parts. Nunan stated, 'If you want to be precise, if you're trying to treat a prostate and you have a bladder right next to it, a rectum right next to it, and you want to get in there and just treat the prostate, you can do that better, more precisely with the linac.'[33] There is no doubt that Kaplan's targeting of malignant lymphomas like Hodgkin's disease was related to his laboratory research on mice. However, the specific advantages of the machine must have been among the crucial elements that moved Kaplan to the treatment of particular lymphomas.

The first clinical trials of Hodgkin's disease (L-1 and L-2 studies from 1962–67) were carried out with 132 patients, mainly those with early clinical stages, Stage I (one single lymphoma) and Stage II (two and more lymphomas), and a few with intermediate Stage III (lymphomas spread to both sides of the diaphragm). Kaplan's prior bio-radiological research led him to the hypothesis that a certain chain of lymph nodes was directly connected with the distribution of Hodgkin's disease, which had long been said to develop in an unpredictable and random way. Adopting new methods of bipedal lymphography, Kaplan and his lifetime collaborator, Saul Rosenberg, successfully showed that far from being random, Hodgkin's disease spread distinctly through lymphatic channels.[34] Performing a series of lapatomies with splenectomies, they found that the spleen was involved in the spread of Hodgkin's disease.[35] These new findings led Kaplan to demonstrate for the first time that a 'Tumoricidal Radiation Dose,' though moderate compared with the earliest experiments, could be feasible and tolerated. Kaplan's group also found it necessary to irradiate the large field encompassing multiple lymph node areas above and below the diaphragm, which they called 'Total Lymphoid Irradiation (TLI)'.

Kaplan's clinical experimentation paid off. Most of Kaplan's articles in the 1960s continued to be based on clinical trials using 6 MeV MLAs. The application of high doses of 3,500 to 4,000 rads for 4 weeks reduced dramatically the relapse rate from 77 percent to 13.3 percent. By 1966, over 90 percent of early stage (Stage I and Stage II) Hodgkin's disease became curable and the survival rate after 5 years increased to

Figure 8.4 Clinical Results in the Treatment of Hodgkin's Disease.

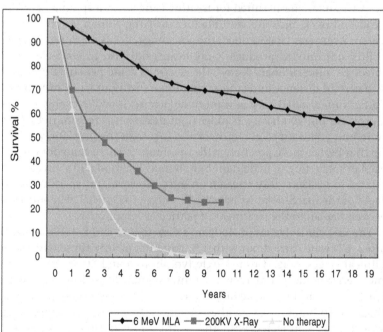

Note: 1) 1718 patients, all stages, treated since 1960 with MeV MLA and/or chemotherapy by Kaplan at Stanford; 2) 285 patients, all stages, treated with high dose 200 KV X-ray therapy at Princess Margaret Hospital; 3) 754 patients, all stages, with no specific therapy (historic).
Source: Kaplan Collection at Lane Medical Library, UODJ3, Series I Box 9.

79 percent. Between 1968 and 1974 Kaplan's group performed clinical trials on 367 patients including those with Stage IV Hodgkin's Disease (neoplasm spread in many places). They not only performed TLI but also combined it with adjuvant multiagent chemotherapy with nitrogen mustard, vincristine, procarbazine, and prednisone (MOPP), a combination which improved greatly both the survival rate and the remission rate of the patients with Stage III and Stage IV Hodgkin's Disease. Kaplan's clinical results represented one of the major advances in cancer therapy in the 1960s.

Kaplan's effort to combine clinical research with radio-biological experiments, then, was little short of revolutionary. Radiotherapy prior to Kaplan's innovations was still quite backward in the United States, compared to Britain or Sweden; it tended to be no more than a palliative technique. In contrast, Kaplan created a successful curative protocol for the MLA. However, once the professional protocol was established, gone were the traditional notions of bedside therapeutics and clinical experience. Discovering a tumor, measuring its needed dosimetry, and pinpointing the radiation began to comprise the entire procedure of medical treatment. Unlike the traditional bedside clinics of the early-twentieth century, where doctors used their expertise to enhance patients' natural healing powers, here emerged a new mode of heroic and interventionist treatment with highly scientific instrumentation.

Conclusion

In this chapter we have shown that Henry Kaplan brought a new paradigm of science-based medicine to Stanford. We have also argued that Kaplan's medical entrepreneurship was based on the building of a new and invasive instrument, the MLA. Kaplan's ambitious vision of transforming medicine at Stanford was made possible by his close collaboration with Edward Ginzton and other physicists and engineers, who saw medical instrumentation as a growth opportunity. In addition to inventing a new medical tool, Kaplan endeavored to establish radiotherapy as a distinct and respected discipline. With the help of Stanford's engineers, Kaplan attempted to establish accelerator-based treatment protocols and worked actively on spreading them to other hospitals and medical centers in the US.

We would like to emphasize that Kaplan's case was representative and paradigmatic of American medicine and medical care in the post-war period. These developments at Stanford mirrored broader changes in many US medical schools. The University of California, Los Angeles is another case in point, as Lenoir and Hays have shown.[36] UCLA owed its dramatic transformation from a regional college into a leading research university to President Robert Sproul and to the entrepreneurial activities of Stafford Warren. Warren, a radiologist, came from the University of Rochester with his entire radiology department and started the medical school at UCLA. Exploiting his close ties with the Atomic Energy Commission, Warren generated a stream of Federal grants and contracts for the UCLA medical school. These medical entrepreneurs were eager to transform medicine into a science-based and interdisciplinary domain integrating the sciences of biology, chemistry, and physics as well as recent developments in engineering. They wished to introduce new and powerful medical technologies into the hospital. Because of these developments, American medicine experienced a drastic transformation in the post-war period. Kaplan's shaping of radiotherapy at Stanford was a typical example of these changes.

Notes

1   There is a growing historical interest in medical technologies, especially in their relations to the medical workplace, the hospital. Key in this field are the works by A.B. Davis, *Medicine and its Technology: An introduction to the history of medical instrumentation* (Westport, Connecticut: Greenwood Press, 1981); S.J. Reiser, *Medicine and the Reign of Technology* (Cambridge: Cambridge University Press, 1981); and J.D. Howell, *Technology in the Hospital: Transforming patient care in the early twentieth century* (Baltimore: Johns Hopkins University Press, 1995). To supplement these works, this chapter will offer new insights on the construction of medical instruments and on the making of contemporary medicine.

2   For this, see A. Cunningham and P. Williams (eds), *The Laboratory Revolution in Medicine* (Cambridge: Cambridge University Press, 1992); L.S. Jacyna, 'The Laboratory and the Clinic: The impact of pathology on surgical diagnosis in the Glasgow Western Infirmary, 1875–1910,' *Bulletin of the History of Medicine*, 62 (1988), 384–406.

3   For the relation between medicine and industrial society, see R.E. Brown, *Rockefeller Medicine Men: Medicine and capitalism in America* (Berkeley: University of California Press, 1979); J. Pickstone, *Medicine and Industrial Society: A history of hospital development in Manchester and its region, 1752–1964* (Manchester: Manchester University Press, 1985); D. Porter, 'Medicine and Industrial Society: Reform, improvement, and professionalization,' *Victorian Studies*, 37 (1993), 129–39.

4   S. Blume, *Insight and Industry: On the dynamics of technological change in medicine* (Cambridge, Mass.: MIT Press, 1992).

5   N. Rosenberg, A.C. Gelijns and H. Dawkins (eds), Committee on Technological Innovation in Medicine, Institute of Medicine, *Sources of Medical Technology: Universities and industry* (Washington, D.C.: National Academy Press, 1995). See also D.A. Holaday, 'Where Does Instrumentation Enter into Medicine?' *Science*, 134 (1961), 1172–7.

6 For a discussion of the rise of science-based and instrumentation-centered medicine, see P. Keating and A. Cambrosio, *Biomedical Platforms: Realigning the normal and the pathological in late-twentieth-century medicine* (Cambridge, Mass.: MIT Press, 2003); J.P. Gaudillère, 'Mapping as Technology: Genes, mutant mice, and biomedical research (1910–1965),' in H.J. Rheinberger and J.P. Gaudillère, *Classical Genetic Research and Its Legacy: The mapping cultures of twentieth-century genetics* (London: Routledge, 2003).

7 Because of the word limit, we can't articulate here the relationship between Kaplan and his group's cancer therapy protocol and the 'War on Cancer' program in the 1970s during which 'cancer' itself underwent a remarkable transformation. See J.H. Fujimura, *Crafting Science: A sociohistory of the quest for the genetics of cancer* (Cambridge, Mass.: Harvard University Press, 1996).

8 See for example, H. Kaplan, 'Observations on Radiation-induced Lymphoid Tumors of Mice,' *Cancer Research*, 7 (1947), 141–7; H. Kaplan, 'Comparative Susceptibility of the Lymphoid Tissues of Strain C57 Black Mice to the Induction of Lymphoid Tumors by Irradiation,' *J. Nat. Cancer Research*, 8 (1948), 191–7.

9 H. Kaplan and M.B. Brown, 'Inhibition by Testoresterone of Radiation-Induced Lymphoid Tumor Development in Intact and Castrate Adult Male Mice,' *Cancer Research*, 11 (1951), 706–8; H. Kaplan, W.H. Carnes, M. Brown and B.B. Hirsch 'Indirect Induction of Lymphomas in Irradiated Mice: I. Tumor incidence and morphology in mice bearing nonirradiated thymic grafts,' *Cancer Research*, 16 (1956), 422–5; H. Kaplan and M. Bagshaw, 'The Stanford Medical Linear Accelerator: III. Application to clinical problems of radiation therapy,' *Stanford Medical Bulletin*, 15 (1957), 141–51.

10 'A conversation with Henry Kaplan,' December 7, 9, and 14, in Henry Kaplan folder, Special Collections in Lane Medical Library, Stanford University.

11 For work in angiocardiography in the radiology department, see H. Abrams and H. Kaplan, *Angiocardiographic Interpretation in Congenital Heart Disease* (Springfield: Thomas, 1956) and H. Abrams, 'An Approach to Biplane Cineandiocardiography, Parts I, II, and III,' *Radiology*, 72 (1959), 441–50, 735–40 and 73 (1959), 531–8.

12 'A conversation with Henry Kaplan'.

13 Radiation therapy on cancer in the post-war period was largely initiated by military funds. With the onset of the Cold War and the Korean War the appropriation of the contract research that funded university research through government organizations proliferated even in the field of bio-medicine. Stanford was one of the universities that most epitomized this trend. For the case of Donnall Thomas and his relation with the Department of Defense and the Atomic Energy Commission, see G. Kutcher, 'Cancer Therapy and Military Cold-War Research: Crossing epistemological and ethical boundaries,' *History Workshop Journal*, 56 (2003), 105–30.

14 For a discussion of radiotherapy from the 1930s to the 1950s, see J. Del Regato, *The Radiological Oncologists: The unfolding of a medical specialty* (Reston, VA: Radiological Centenniel, 1993).

15 'Conversations with Kaplan,' Kaplan Papers, Series VII, Box 1, Stanford Special Collections in Green Library.

16 For institutional changes in the biomedical field at Stanford, see Eric J. Vettel, 'The Protean Nature of Stanford University's Biological Sciences,

1946–1972,' *Historical Studies in the Physical and Biological Sciences*, 35 (2004), 95–113.

17 M. Shultz, 'The Supervoltage Story, Janeway Lecture, 1974,' *The American Journal of Roentgenology*, 124 (1975), 541–59.

18 In the early-1950s, HVEC did a modest business selling 2 MeV Van de Graaff machines to hospitals. 'High Voltage Engineering Corporation, Annual Report, 1954,' 'Balance Sheet, 1954,' 'The Story of High Voltage Engineering Corporation,' Ginzton Papers, Box 11, Stanford Special Collections in Green Library.

19 'A Conversation with Henry S. Kaplan: Interviews Conducted December 7, 8, 9, 1983,' Kaplan papers, Series VIII, Box 1, p. 24, Stanford Special Collections in Green Library.

20 'A conversation with Henry Kaplan'. On radiation therapy in Europe, see Patrice Pinell, 'Cancer,' in: R. Cooter and J. Pickstone (eds), *Medicine in the Twentieth Century*, Amsterdam: Harwood Academic, 2000, 671–86.

21 The hospitals and manufacturers Ginzton visited in the UK were Hammersmith Hospital, Christie Hospital and Holt Radium Institute, Victoria Infirmary, Brown Boveri and Co., and Metropolitan Vickers Electrical Co. LTD. Kaplan Papers, Series I, Box 5.

22 Kaplan Papers, Series VII, Box 1, Box 3, Ginzton Papers, Box 6, 'Medical Accelerator,' Stanford Special Collections in Green Library.

23 Ginzton Interview, number IV, conducted by H. Lowood, Ginzton Papers, ACCN 1991–13 Box 21, Stanford Special Collections in Green Library.

24 E.L. Ginzton, 'The Medical Linear Accelerator,' *The Varian Associates Magazine* (June 1973), 27, Varian Papers, Series: Varian Associates, Box 9, folder 4, Stanford Special Collections in Green Library.

25 E.L. Ginzton, K.B. Mallory, and H. Kaplan, 'The Stanford Medical Linear Accelerator: I. Design and development,' *Stanford Medical Bulletin*, 15 (1957), 123–40; Kaplan and Bagshaw, 'The Stanford Medical Linear Accelerator: III.'

26 C. Beyers, 'The Medical Linear Accelerator: A Joint Venture,' Varian's documents, Box 9.1a, Special Collections in Lane Medical Library.

27 For a discussion of klystron technology, see Christophe Lécuyer, *Making Silicon Valley: Innovation and the growth of high tech, 1930–1970* (Cambridge, Mass.: MIT Press, 2006).

28 Stanford Oral History Project, Interview with Ed Ginzton, Fourth Interview, 23 March 1988, conducted by Henry Lowood, in Ginzton Papers, ACCN 1991-113, Interview with Ginzton IV, pp. 5–6, Stanford Special Collections in Green Library.

29 M. Weisbluth, C.J. Karzmark, R.E. Steele, and A. Selby, 'The Stanford Medical Linear Accelerator. II. Installation and physical measurements,' *Radiology*, 72 (1959), 242–53; P.R. Steed, 'The Stanford Medical Linear Accelerator. IV. Patient dosimetry,' *Stanford Medical Bulletin*, 15 (1957), 152–8.

30 C.J. Karzmark, 'Large-field Superficial Electron Therapy with Linear Accerelator,' *British Journal of Radiology*, 37 (1964), 302–5; C.J. Karzmark, R. Loevinger, and M. Weissbluth, 'A Technique for Large-Field, Superficial Electron Therapy,' *Radiology*, 74 (1960), 633–44.

31 H. Kaplan and M.A. Bagshaw, 'The Stanford Medical Linear Accelerator: III'; H. Kaplan, 'The Radical Radiotherapy of Regionally Localized Hodgkin's

Disease,' *Radiology*, 78 (1962), 553–61; H. Kaplan and S. Rosenberg, 'Extended-field Radical Radiotherapy in Advanced Hodgkin's Disease: Short-term Results of 2 Randomized Clinical Trials,' *Cancer Research*, 26 (1966), 1268–76.

32 Stanford Medical Center News Bureau, 'Cancer Research Center Established,' August 30, 1961, Special Collection in Lane Medical Library, Archives UODJ3, Box 9.17. H. Kaplan, 'Long-term Results of Palliative and Radical Radiotherapy of Hodgkin's Disease,' *Cancer Research*, 26 (1966), 1250–2.

33 An Interview with Graig S. Nunan, conducted by Sharon Mercer, Varian Associates Inc. Oral History Project, Ginzton Papers, ACCN 1991-133, Box 21, p. 20, Stanford Special Collections in Green Library.

34 S.A. Rosenberg and H. Kaplan, 'Evidence for an Orderly Progression in the Spread of Hodgkin's Disease,' *Cancer Research*, 26 (1966), 1225–31.

35 E. Gladstein, J.M. Guernsey, S.A. Rosenberg, and H. Kaplan, 'Value of Laparotomy and Splenectomy in the Staging of Hodgkin's Disease,' *Cancer*, 24 (1969), 709–18; E. Gladstein, H.W. Trueblood, L.P. Enright, S.A. Rosenberg and H. Kaplan, 'Surgical Staging of Abdominal Involvement in Unselected Patients with Hodgkin's Disease,' *Radiology*, 97 (1970), 425–32.

36 T. Lenoir and M. Hays, 'The Manhattan Project for Biomedicine,' in Phillip Sloan (ed.), *Controlling Our Destinies* (South Bend: University of Notre Dame Press, 2000).

9
Hexamethonium, Hypertension and Pharmaceutical Innovation: The Transformation of an Experimental Drug in Post-war Britain

Carsten Timmermann

Introduction

Hypertension is one of the most common medical problems in the developed world.[1] It is also one of those modern conditions that we do not find discussed in old medical textbooks, and hypertension has not always been self-evidently viewed as a disease.[2] High blood pressure has variously been interpreted, for example, as a physiological response to ageing, a metaphor for life under the strains of modernity, or the symptom of a so far unrecognized underlying genetic disease. Only in the last five decades it has become acceptable to treat high blood pressure, the symptom, without knowing the causes. Thanks to a new measuring technology (the familiar combination of an inflatable cuff, a sphygmomanometer and a stethoscope), blood pressure could be measured easily outside the laboratory since the early-twentieth century, but high blood pressure was not perceived as a public health problem and the method was initially not taken up widely by physicians.[3] In Britain such reluctance prevailed until after World War II, as part of a general scepticism towards using laboratory methods in the clinic.[4] In the US, in contrast, medical directors in the life insurance industry were quick to realize the potential of using this portable technology extensively, leading to the drawing up of tables that associated blood pressure with insurance risks. Their calculations made blood pressure interesting to epidemiologists and visible to a wider audience of physicians.[5] In Britain, hypertension gained profile as an issue of medical interest immediately after World War II. This coincided with the intro-

156

duction of the National Health Service (NHS) in 1948. More importantly, though, the post-war years also saw a strengthened and proactive Medical Research Council (MRC) that sought to set new
standards for clinical research, a re-orientation in the pharmaceutical
industry with a stronger focus on research and development, and not
least the development of the first drugs that targeted high blood pressure.[6] These new drugs and the constellation around the introduction
of new treatment methods for hypertension, roughly between 1948
and 1960, coinciding with what has often been described as a 'golden
age' of modern medicine, are the subject of this chapter.[7]

I will explore how hypertension was transformed into a treatable
disease when a new class of drugs, the ganglion blockers, moved from
the laboratory to the bedside and eventually turned from experimental tools into routine treatment, by looking at the roles played by
researchers in laboratories and clinics, and the debates over practical
(and to some degree ethical) issues of the long-term administration of
antihypertensive drugs. The sources from which I am reconstructing
this story (mostly files from the archives of the MRC) also allow me
some more general statements about the dynamics of biomedicine in
Britain and the interactions between the public sector and the drug
companies in this crucial period. The ganglion blockers are among the
fruits of the MRC's intense campaign for the establishment of a new
form of academic medicine in the country. Pharmaceutical companies, as we will see, initially played subordinate roles, while the
MRC pulled the strings.[8] Whereas in the United States, as Nicolas
Rasmussen has shown, drug companies actively approached and
recruited medical academics, the story I will tell is one of 'shy' drug
company representatives, recruited by the MRC for its plans.[9] Only in
the mid-1950s, when consensus was emerging that some forms of
hypertension benefited from long-term drug treatment, as Viviane
Quirke has shown for ICI, pharmaceutical companies became
more proactive in the search for new antihypertensive drugs.[10] Hexamethonium and other ganglion blockers prepared this consensus. But
the first responses to proposals to use hexamethonium as a routine
drug were cautious, and I will demonstrate how important the development of an effective system for managing its effects, both desired
and undesired, with nurse-technicians and patients playing important
parts alongside clinicians, was for the successful transformation of the
drug. This clinical setup for the treatment of high blood pressure with
hexamethonium was devised, as we will see, by a British clinician with
strong MRC links, but on the other side of the globe, in a teaching

hospital in New Zealand. For the first stage of our journey following the ganglion blockers from the laboratory to the clinic, however, we are now turning to the physiology laboratories of Cambridge, England, and Cambridge, Massachusetts.

Quaternary ammonium salts: from the physiology lab into the medicine cabinet

In the 1940s, the quaternary ammonium salts soon to be known as ganglion blockers were longstanding tools for physiological research. Pharmacologists and physiologists, notably the members of the Cambridge school of physiology had been interested in these drugs because some of their effects on experimental animals resembled those of the nerve poison curare. Their main focus of interest was not the clinical application of the drugs; they were studying the molecular mechanisms of the nervous system. The blood pressure lowering effect of some of these compounds, especially tetraethylammonium (TEA) had been noted on a few occasions by researchers, but none of them had seen it as particularly significant.[11] The name, ganglion blocking drugs, and the clinical potential of these compounds was a product of academic research on TEA by two young physiologists, George Acheson and Gordon Moe at Harvard Medical School in 1945 and 1946. Prompted by a pre-war article, Moe and Acheson purchased a sample of the bromide salt of TEA from Eastman Kodak, and found that its effects in an animal experiment were pharmacologically interesting. To explain these effects, Moe suggested that the TEA ion blocked the ganglia of the autonomic nervous system, the wiring, as it were, that controlled the unconscious, regulatory functions of the body.

So far our drug has not left the laboratory. But what distinguished Acheson and Moe from the researchers who had previously used quaternary ammonium compounds was the fact that they found it worthwhile to explore possible clinical applications for TEA. While some of the earlier researchers suggested there might be clinical uses for these drugs, none had considered the long-term treatment of high blood pressure a worthwhile application. 'The major ideas that Gordon Moe and I arrived at in 1943 and 1944 were already in the literature. Why then,' George Acheson asked in 1975, 'did our work make a splash whereas theirs did not?'[12] To Acheson, what made the difference was the fact that during World War II the physiology lab and the clinic had become far more closely linked, and that researchers had lost their reservations against talking to the pharmaceutical industry. Moe and

Acheson collaborated both with clinicians and with industry and Acheson explicitly credits the post-war context for this change. Drugs were gaining a better status among doctors, as a result of the success of sulfonamides and penicillin, he remembers, and drug companies had become 'more respectable'.[13]

Acheson and Moe initiated clinical tests for TEA, but they were not very successful. The effects of the drug did not last long enough and, while the tests established the notion of ganglion blockade as a possibility, TEA was not the drug that could move smoothly from bench to bedside. The first ganglion blockers that were to prove clinically useful were the methonium compounds, and for this part of the story we are now moving to Britain. As was the case for TEA, the observation that methonium compounds had an effect on blood pressure was partly a serendipitous discovery, a by-product of work that had its roots in the Cambridge physiological-pharmacological tradition, combined with a heightened sensitivity for cardiovascular effects in the post-war years. That hexamethonium was going to turn into a high blood pressure drug, though, was by no means obvious when the physiologist William Paton came across the methonium compounds. He was working at the time in Henry Dale's old laboratory at the MRC's National Institute for Medical Research (NIMR) at Hampstead, where researchers were just about to revert from war research to peace-time activities. Paton himself was about to turn into a respiratory physiologist.[14] Paton, son of a clergyman, had studied physiology at Oxford and like many other pioneers of medical research at this time, studied medicine at University College Hospital (UCH) Medical School, where he came into contact with the MRC style of medical research.[15] He joined the staff at the NIMR in 1944. In 1952 he moved to UCH for two years as Reader, a joint-appointment between Rosenheim's Medical Unit and the Department of Pharmacology. From 1954 to 1959 he held a professorship of pharmacology at the Royal College of Surgeons, before succeeding J.H. Burn on the Oxford chair of Pharmacology. Hexamethonium entered the scene when Paton's laboratory at the NIMR – not for the first time – took on some work for the Institute's Division of Biological Standards. They were asked to look at the toxicity of a promising antibiotic substance, lichenoformin that a member of the Chemotherapy Division, R.K. Callow, had isolated from cultures of *Bacillus licheniformis*. Injecting the substance into an anaesthetized cat Paton found that nothing happened for about 25 seconds, when he measured an abrupt and transient fall in blood pressure. The substance seemed to trigger the release of a vasodilator (a substance that widens blood vessels), most likely histamine, in a way that impressed Paton and his colleagues because 'it was such an interesting and clean response.'[16] The excitement over clean and pure responses was typical among pharmacologists and indicative of their ideal of reducing a drug's action, if possible, to one active principle, the interaction between a chemical structure and a physiological function. Clinical usefulness, however, as we will see later, depended on more than just understanding the mechanism of a drug.[17]

Callow told Paton and his colleagues more about the chemical structure of lichenoformin, and they went upstairs to the NIMR's Chemistry Division under Harold King, where they obtained a range of substances with related chemical groups, in many of whom they found that they also released histamine and triggered a similar 'delayed repressor response'. Such responses had been observed before, but according to Paton 'it was entrancing to see this simple, specific response, to detect a coherent pattern emerging in the relationship between chemical structure and pharmacological response, and to see possible new approaches to the physiology of histamine.'[18] The series of compounds they tested also contained two derivatives of the methonium series, C8 and C16 (the figure stands for the number of atoms in the carbon chain between the two quaternary ammonium groups). C8 produced a somewhat different, but also remarkably clean response. It proved to be a potent neuromuscular blocking agent, a drug that led to temporary paralysis, with potential applications for anaesthesia and surgery.

The decision to further pursue work on this group of substances was informed by the growing interest in developing synthetic drugs with curare-like action, to treat convulsions and as muscle relaxants for surgery. King had long been researching the active principle of curare. A significant factor was the arrival of Paton's co-worker, Eleanor Zaimis, who was experienced in chemistry as well as pharmacology. Encouraged by King, Paton and Zaimis decided to undertake a study of the whole series of methonium compounds and their effects, with carbon chain lengths from C2 to C12. Around the time when they started this work, Paton and Zaimis discovered that Ing and Barlow at Oxford were studying the same compounds, in a more systematic way and as one among a number of other homologous series, in an attempt to establish the structure-action relationships of neuromuscular block.[19] In 1948, both groups arranged a simultaneous publication (as Letters to the Editor) in the journal *Nature*. It is noteworthy that effects on blood pressure were mentioned in neither paper.[20] Such effects had been noticed, however. Paton and his colleagues injected the drug into a number of laboratory animals, which revealed that C5 and C6, the compounds most active in anaesthetized cats, had significant effects on the animals' blood pressures.[21] Injected into a rabbit's ears, they made the ears flush in a bright red. According to his reminiscences (published in 1982), this reminded Paton of a lecture he had attended as an undergraduate on Claude Bernard's experiments on the effects of interrupting the sympathetic innervation to the rabbit's ear, pointing towards the sympathetic ganglia as a site of action of C5 and C6. This may well be a retrospectively constructed memory, but it illustrates Paton's interpretation of the ganglion blockers as the culmination of a longstanding, great tradition in physiology. Again, Paton marvelled over the 'very clean action' of the drug.[22] 'The analysis of hexamethonium's action,' he remembers, 'was a most enjoyable experience.'[23]

There was some interest in the drug's possible use as anti-hypertensive, mostly due to Acheson and Moe's series of articles two years earlier on TEA and its effect on blood pressure and an increasingly lively debate over the aetiology and treatment of hypertension. In his reminiscences Paton remembers that: 'At all events with this, and with the experience of sympathectomy, the possibility that hexamethonium might be usable in hypertension as a clean, specific ganglion blocker was one to consider.'[24] At the time, however, the methonium compounds to Paton and Zaimis were above all interesting pharmacological research tools. Only in a further letter to *Nature* in November 1948 they suggested – in one brief sentence at the end of the paper – that C6 'offers possibilities of clinical usefulness in such fields as hypertension and vascular disease, whenever tetraethylammonium iodide has too brief or slight an action.' In the following section we will see what it took to establish uses for the clean, specific responses that Paton observed in his laboratory in the much messier world of the clinic. Here too, the MRC and the networks that the Council created to promote its model of medical research played decisive roles.

Clinical trials and invisible industrialists

The MRC appointed Paton to chair an informal committee investigating possible clinical uses of the methonium drugs.[25] Along with Frank Green, the Council's Chief Medical Officer, Paton was the main correspondent in all matters concerning clinical uses of the methonium compounds (he also prepared the official report on the drugs).[26] Not long after Paton and Zaimis heroically tested the effects of several methonium compounds on themselves at Westminster Hospital, together with the anaesthetist Geoffrey Organe, the MRC initiated the first clinical study of pentamethonium for the treatment of hypertension.[27] The Lancet published the results in August 1948.[28] The investigation was carried out, 'at the request of the Medical Research Council', by P. Arnold and Max Rosenberg of the Medical Unit at University College Hospital, where Rosenberg had succeeded Thomas R. Elliott on what was the first, MRC-funded, full-time chairs for clinical research in Britain.[29] They reported that 'pentamethonium iodide has an action similar to that of tetraethylammonium chloride, but it is effective in smaller doses' and that 'an excessive fall in blood-pressure has been the only serious toxic effect so far observed'.[30] Anaesthetists experimented with hexamethonium as an antidote to the muscle relaxant decamethonium and viewed blood pressure reduction as an undesired side effect: 'In the very first case it seemed that the ensuing hypotension might be severe enough to be fatal.'[31] It is not obvious from these early publications that soon the lowering of blood pressure would be discussed as the main rather than a dangerous side effect of ganglion blockers. During an MRC Conference on Clinical Tests of Methonium Drugs on 22 June 1950, hypertension was one among several possible indications including use in anaesthesia and in the treatment of stomach ulcers.[32]

The initial responses to the methonium compounds and the suggestion that they might provide a useful treatment option for high blood pressure were cautious. Increasingly, though, they were cautiously optimistic. A series of articles reporting results of studies, involving rather small numbers of patients for today's standards, were published in the *Lancet* in 1950. So were reports on other antihypertensive treatments, all relatively new, such as the Kempner Rice Diet or a variety of surgical procedures. Some authors characterized the effect of hexamethonium as 'medical sympathectomy'.[33] Sympathectomy was an irreversible surgical intervention where connections in the sympathetic nervous system of patients with malignant hypertension were severed.[34]

Representative of the early, cautiously positive responses was that by Stephanie Saville, reporting on a study undertaken at St Martin's Hospital, Bath, in a well-established clinic for hypertension, where the customary treatment over years had been rest in bed, sedation, and regular venesection, to which 'many patients failed to respond significantly or to maintain initial improvement'. Five cases of malignant hypertension were experimentally treated with pentamethonium, starting in December 1949, initially in the hospital and later also as outpatients. The aim was to lower the blood pressure enough to relieve symptoms without provoking others due to hypotension. While Saville was reluctant to predict the final outcome, she observed that 'none of the five cases sufficiently treated to date has failed to obtain relief from symptoms'. Saville's conclusion: 'The results in 5 patients suggest that pentamethonium bromide may be useful in the treatment of at least some forms of hypertension.'[35] Allan Campbell and Eric Robertson at the Royal Alexandra Infirmary, Paisley found (in a study with eight patients) that 'hexamethonium seems to provide a useful method of reducing the blood pressure in severe hypertension with ready administration and relative freedom from toxicity'.[36] The main problem was finding the right dosage, and some of the side effects were quite drastic, too. R. Turner in Edinburgh suggested that 'methonium drugs have as yet no place in the routine management of patients, though they may prove useful in the treatment of resistant symptoms related to hypertension. We need more information about their precise action, and for the present it will be most profitable to study, in detail, patients who might otherwise be treated by sympathectomy.'[37]

These clinical studies on the methonium compounds were initiated by the MRC and not, as we would expect today and as even some insiders assumed, by the pharmaceutical industry.[38] Austin Doyle, a colleague of Smirk projects today's patterns of drug development on the past in his reminiscences, suggesting that May & Baker approached clinicians with the request to test the methonium compounds in the clinic.[39] The files in the National Archives tell a different story, one where the MRC pulled the strings in the crucial period between 1948 and 1952, with Paton as unofficial patron of the ganglion blockers, supported and backed by Frank Green. The ganglion blockers provide us with an example for a major pharmaceutical innovation that had its origins in the public sector, with the industry playing a subordinate role. The Council mediated between the clinical researchers and the companies that provided the drugs, which included Burroughs Wellcome, Allen & Hanbury's, and more often than the others, May & Baker. Like TEA, the methonium compounds were not patent-protected, and the drugs from different producers were different only with regard to preparation and sometimes purity. The MRC acted as a booster for the methonium drugs and a 'matchmaker' between clinical researchers and drug manufacturers, who assumed, it seems, that it was improper for them to approach the researchers directly. A memo by Green in 1950 illustrates the mechanics of this interaction and the central role of the Council:

> I telephoned to Dr Forgan [of May & Baker] to ask him to send supplies of pentamethonium and hexamethonium for trial by the mouth in cases of hypertension. Forgan said he would be glad to do

so. He asked whether it would be in order for him to invite Rosenheim to express an opinion on the relative effects of these two substances, as May & Baker have the impression that hexamethonium is so much more useful than pentamethonium that the latter may be on its way out. I said that there would be no objection whatever from our point of view to his asking Rosenheim this or any other question which occurred to him. (Note: May & Bakers still seem to be under the erroneous impression that the Council, when they arrange clinical trials, prohibit the manufacturers from communicating directly with the investigator; this arrangement applied in the early days of the Therapeutic Trials Committee, but it has long since been abandoned as both unpopular and inefficient.)[40]

In the same memo, Green noted that he had suggested to Forgan that May & Baker should look into the possible uses of the methonium drugs in ulcerative colitis, and proposed to contact a number of suitable clinical investigators for such questions and organize a conference, with Paton as secretary. On a different occasion Green wrote to Paton that:

> [W]e should certainly not regard it as improper for the firm of Geigy – or, indeed, any other firm – to approach Kay and Smith direct on a scientific question such as you mention. Of course, there is no obligation on Kay and Smith to assist Geigy with advice unless they like to do so.[41]

The power relationships, it seems, were distinctly different from those that Nick Rasmussen has described for the United States, where he finds patterns of company-funded clinical research very similar to today's as early as the inter-war period.[42]

Therapeutic enthusiasm

Although the MRC sought to promote the ganglion blockers, the initial responses from clinicians involved in the early clinical trials were cautious. In order to understand the transformation of hexamethonium from experimental drug to routine treatment for malignant hypertension, in this section we will look at the role of clinicians such as Frederick Horace Smirk, who we might want to call 'therapeutic enthusiasts'.[43] Smirk may seem a somewhat marginal figure at the first glance. Looking closer at his career, though, it becomes clear that his department at University of Otago Medical School in Dunedin, New Zealand was modelled on the MRC-funded units run by T.R. Elliot and Thomas Lewis at University College Hospital (UCH) London in the inter-war years. 'The clinic', Smirk stated in a report on his hypertension clinic in Dunedin, 'has served both research and routine requirements and often there has been no distinction between these requirements'.[44] A product of the work there was, besides a routine method of treating hypertension with drugs, a major textbook on *High Arterial Pressure*.[45] Dunedin became a satellite of clinical research in the British metropolis, and in this was not so different from British provincial universities.

Smirk, the son of a Lancashire schoolmaster, had attended Manchester University in the early-1920s, graduating MB, ChB with first-class honors in 1925. He acquired his MD in 1927. After junior posts at Manchester Royal Infirmary, a Dickenson Travelling Scholarship from the MRC allowed him to go to Vienna and a Beit Memorial Medical Research Fellowship to join T.R. Elliot's unit at UCH. As Smirk's Dunedin colleague F.N. Fastier put it, UCH 'was an almost inevitable choice for tenure of the Beit Fellowship. No other British medical school could have provided during the 1930s a more stimulating environment for those "new men" who were intent on fusing the skills of the laboratory and the clinic.'[46] The 'new men' who joined the research staff at UCH with a Beit fellowship included eminent figures such as George Pickering or John McMichael, who would play an important role later in Smirk's career and in the hexamethonium story.[47]

While many of his contemporaries found positions in Britain, often full-time posts in the new research units established by the MRC, Smirk's next move after teaching appointments in the departments of pharmacology and medicine at UCH was to the chair of pharmacology at Cairo University in 1935. He stayed in Cairo for four years, during which he turned to the search for drugs that had an effect on blood pressure, screening nearly 1,500 commercially available chemicals in stray dogs, with little success. In 1940 he was appointed to the first full-time chair of medicine at Otago, replacing the part-time professors of systematic and clinical medicine who had retired the previous year. The move towards full-time professors who were expected to do research seemed to parallel what was happening in Britain. Later Smirk was able to link Dunedin to the metropolis in other ways. He was a scientific entrepreneur who managed to secure funding for his work from a variety of sources, including pharmaceutical companies. This success was undoubtedly due to his great enthusiasm for pharmaceutical solutions and his success in turning the ganglion blockers into routine drugs. He was also credited with the development, by selective inbreeding, of laboratory rats with inherited hypertension.[48] The greatest coup of his career as research administrator, however, was to secure £120,000 from the Wellcome Trust for a completely new clinical research institute in Dunedin in 1960, then among the biggest grants the Trust had ever made. This was undoubtedly helped by the fact that McMichael by then was a Wellcome Trustee and Green the Trust's Medical Secretary.

Initially, however, the war years did not make it easy to establish a functioning laboratory in Dunedin. There was hardly any equipment and much had to be improvised, supplies were unreliable, and there was not much space for Smirk and his staff. Furthermore, Smirk was faced with a heavy teaching and administrative load. Like Paton's at the NIMR, Smirk's research turned to typical war-time tasks, such as the study of nitrogen mustards and the search for a pharmaceutical treatment for shock suitable for situations where quick fluid replacements were not possible. After the end of the war he returned to the search for antihypertensive drugs. This search received a boost when in 1949 he spent a sabbatical at the Postgraduate Medical School at Hammersmith Hospital in London, where McMichael was director and professor of medicine, the year when Paton, Zaimis and Organe, and later, more importantly, Arnold and Rosenheim, Elliot's successor at UCH, published the first results of their experiments with pentamethonium and hexamethonium in humans.[49] Smirk, who received supplies of the drugs from May and Baker when he returned to New Zealand, it seems, had found the compounds he was looking for.

An important problem of the methonium compounds which accounted for much of the scepticism by clinicians who had tested them was their low solubility in water and the resulting low (and unreliable) rate with which the drugs were absorbed into the bloodstream when taken by mouth. Smirk overcame this by devising a regimen of subcutaneous injection, a management solution very similar to the administration of insulin to diabetics. Patients had to be 'titrated', the right dose found for each individual patient, and this dose adjusted as patient bodies got used to the drug. As the Hull physician Edmond Murphy (who characterized Smirk's attitude towards hexamethonium as 'enthusiastic') observed, 'Smirk and Alstad [one of Smirk's co-workers in Dunedin] claimed that with adequate parenteral doses – each patient being a law unto himself – a lowered tension can be maintained indefinitely'.[50] This went along with a strict monitoring regime. Austin Doyle, who joined Smirk's department in 1952, remembers:

> Having been working in a traditional way in a British hospital, I was amazed at the confident and routine way that these difficult drugs were being used. ... Patients would arrive at the clinic at about 8:30 AM, have blood pressures recorded while seated and standing, and be given a dose of hexamethonium subcutaneously by the nurse. Blood pressures were then recorded in both postures at 30-minute intervals throughout the day. At about 12:30 PM Smirk would visit the clinic, look at the data, and order the appropriate afternoon doses. Patients would attend daily until the correct dose had been attained and would then be allowed to leave, having been supplied with tuberculin syringes, needles, and multidose containers of the drug, which they had been trained to use.[51]

The multidose containers had been prepared at the university's pharmacology department by dissolving bulk supplies provided by May and Baker. Some years later, such multidose vials were available prepacked, directly from the drug producers.

Drugs and discipline

Hexamethonium, it seems, disciplined doctors, patients, carers, and technicians alike. 'It is most important', Smirk stated in his account of the practices in the hypertension clinic, 'that the patient should understand something of the working of the drug, and in our experience the technician is invaluable in educating patients, and any special points she cannot answer can be dealt with when the doctor comes round'.[52] He was passionate enough about the importance of his technicians that a passage referring to the women is capitalized in his report:

> THERE IS ONE ESSENTIAL WITHOUT WHICH OUR CLINIC COULD NOT FUNCTION EFFECTIVELY, AND THIS IS THE PRESENCE OF RELIABLE TECHNICIANS INTERESTED IN THEIR WORK AND WITH SUFFICIENT STABILITY, FRIENDLINESS AND POISE TO BE TRUSTED BY THE PATIENTS. WITH INTELLIGENT GIRLS WHO ARE KEEN TO LEARN WE HAVE EXPERIENCED NO DIFFICULTY WHATSOEVER IN OBTAINING FROM THEM ACCURATE PRESSURE READINGS AND COMMENTS ON CORRESPONDING SYMPTOMS, IF ANY.[53]

Smirk devised a treatment manual for patients and carers that explained the reasons for the treatment, suggested ways of dealing with the possible side effects of the drugs, which would disappear soon if patients cooperated and played active parts in the therapy, and warned of the possible consequences of non-compliance or interruption of the treatment.[54] Smirk explained patients that '[t]he object of treatment is to decrease the risk of complications of high blood pressure such as heart failure and stroke, and secondly to relieve symptoms such as headache and breathlessness in cases where these are due to high blood pressure'.[55] Patients were encouraged to perform simple tests on themselves, which relied on them developing an awareness of the effects of the drug on their bodies. The 'standing test' took advantage of one of the most common side effects of hexamethonium, postural hypotension, a sudden lowering of the blood pressure when patients stood up, which led to dizziness. 'Almost all of our well-trained patients can tell the doctor from their own subjective sensations whether a given dose is producing a considerable fall of the blood pressure', as Smirk reports. 'Intelligent and cooperative patients have been entrusted with the fine adjustment of their dose'.[56] The ways in which Smirk's practices dealt with postural hypotension and made use of this side effect, even more than the issue of resorption, indicate that even a difficult drug could be managed.

Smirk blamed the bad results that other clinical researchers reported with the methonium drugs, as Green wrote in a letter to Paton, on '"faulty technique", namely lack of proper ancillary care of the patients'.[57] Smirk's practices involved patients as well as technicians and nurses in active and responsible roles. Tending to the psychology of the patients was part of the regime: 'It is well to remember that nervous tension, worry, quarrels, excitement and adverse emotion lead to elevation of the blood pressure', Smirk wrote. 'It is a part of treatment to lessen such troubles.'[58] How was this done? 'Reassurance by doctor and technician that the milder side effects are in no sense dangerous and will probably disappear anyhow usually leads to a happier frame of mind.'[59] As important, it seems, was the company of other patients. Patients would sit together in groups of four while undergoing their lengthy tests, exchanging experiences, and there was 'something of the atmosphere of a club about the clinic'.[60] Not all were happy, though: 'It is well to realise ... that a drug which, in a proportion of patients, causes side effects will get the blame for all sorts of incidental illness.'[61] Frank Green, after his visit in New Zealand, reported to Paton:

> I talked to some of his patients under treatment in the wards, and I
> think it is fair to say that I got the impression that their enthusiasm

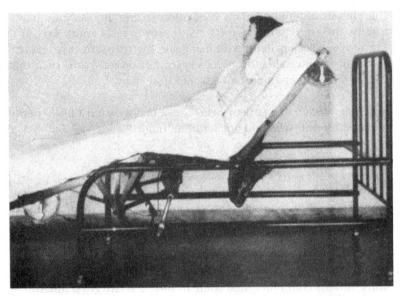

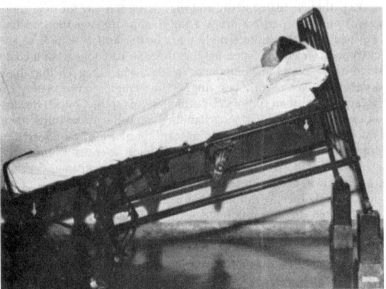

Figure 9.1 Hospital beds raised to compensate for postural hypotension. These beds are examples of the simple solutions used in Smirk's clinic for the management of side effects. Figures from F.H. Smirk, *The Treatment of High Blood Pressure* (Edinburgh: Royal College of Physicians, 1956). The author thanks the Royal College of Physicians of Edinburgh for the permission to reproduce these images.

for hexamethonium therapy is not quite so great as his. Not unnaturally, they dislike the side effects. If, however, as Smirk says, it is possible to keep patients with malignant hypertension alive and reasonably comfortable for periods at least of several years, then that does represent an important advance in therapy.[62]

To be fair, though, the patients who Green met were most likely newly admitted patients who had not yet been 'trained'.

Conclusion

A *Lancet* editorial in 1951 supports Green's conclusions, stating that: 'Early reports leave no doubt that the methonium compounds are the most powerful hypotensive agents yet developed.'[63] Other therapeutic enthusiasts, like Edward Freis were quick to acknowledge this.[64] By the end of the 1950s, hypertension was transformed into a treatable disease. The ganglion blockers, along with the routines designed by Smirk and his colleagues, had come to form a relatively standardized package that circulated easily and that paved the way for new, more specific antihypertensive drugs. Drug therapy for hypertension had returned from the more remote parts of the British Empire to the metropolis. Green and Paton felt that the MRC had done what it could do for the methonium compounds in 1952, when they felt that these drugs no longer needed boosting.[65] Their emphasis was clearly on research rather than commercial exploitation; for the Council hexamethonium was part of a long-standing campaign to establish their concept of biomedicine as the dominant form of medical research in Britain, in the clinic as well as the preclinical sciences. Commercial exploitation followed, however, with several drug companies developing their own, patent-protected ganglion blockers on the back of the success with the methonium compounds. Examples are May and Baker's Gaplegin, or Pendiomide, which was developed and marketed by both Ciba and ICI more or less simultaneously. The new ganglion blockers were not necessarily better antihypertensives than hexamethonium, but they were more easily absorbed by the gut and therefore more reliable when taken by mouth, and therefore disposed with the need for elaborate clinical setups like that developed by Smirk.[66] For several producers, the ganglion blockers provided stepping stones into the new field of pharmaceuticals for blood pressure control.[67]

Modern drug research, we assume, aims to find drugs for specific targets. In the case of hypertension the target was defined and refined as new drugs became available. Drugs like the ganglion blockers and other medicines for chronic 'disorders' have also come to redefine our lifeworld, and particularly notions of what we consider as normal or as pathological. Hypertension in the late-1940s was not what it is today. Malignant hypertension, seriously increased blood pressure which led to life-threatening, pathological changes, today has more or less disappeared from the industrialized world. Starting with the ganglion blockers, new drugs turned hypertension from an acute into a chronic disease that could be managed (rather than cured) and allowed sufferers to lead reasonably normal lives (in a way that was similar to the role that insulin played for diabetic patients). But more than that, the ganglion blockers paved the way for a new therapeutic enthusiasm. In the case of hypertension it has become acceptable, even recommended, to treat a symptomless physiological phenomenon, as if it were a disease.

Acknowledgments

Many friends and colleagues have commented on this paper or earlier versions during its long period of gestation, including panel members and audiences at workshops in London, Durham and Manchester, a conference in Vienna, and a seminar in London. I especially thank my Manchester colleagues Emm Barnes, John Pickstone and Elizabeth Toon for their valuable suggestions for the final version. My research was generously supported by a Wellcome Trust fellowship.

Notes

1 T.E. Kottke, R.J. Stroebel, R.S. Hofman, 'JNC 7 – It's More than High Blood Pressure,' *Journal of the American Medical Association*, 289 (2003), 2573–5.

2 Cf. C. Timmermann, 'To Treat or Not to Treat: Drug research and the changing nature of essential hypertension,' in T. Schlich and U. Tröhler (eds), *The Risks of Medical Innovation: Risk perception and assessment in historical context* (London: Routledge, 2006); idem, 'A Matter of Degree: The Normalisation of Hypertension, circa 1940–2000,' in W. Ernst (ed.), *Histories of the Normal and the Abnormal* (London: Routledge, 2006); I.H. Page, *Hypertension Research: A memoir 1920–1960* (New York: Pergamon Press, 1988); G.W. Pickering, *High Blood Pressure* (London: Churchill, 1955).

3 Cf N. Postel-Vinay, *A Century of Arterial Hypertension 1896–1996* (Chichester: Wiley, 1996); K. Ilsley, *An Historical Study of Arterial Blood Pressure and its Management, from the Seventeenth Century to Modern Times* (MA dissertation: University of Wales Swansea, 2005).

4 Cf. C. Lawrence, 'Still Incommunicable: Clinical holists and medical knowledge in interwar Britain,' in C. Lawrence and G. Weisz (eds), *Greater than the Parts. Holism in biomedicine 1920–1950* (New York & Oxford: Oxford University Press, 1998), 94–111.

5 See, for example, D. Riesman, 'High Blood Pressure and Longevity,' *Journal of the American Medical Association*, 96 (1931), 1105–11. As Rothstein, Davis and others have shown, the life insurance industry in the US assumed many of the functions of public health authorities in Europe. W.G. Rothstein, *Public Health and the Risk Factor: A history of an uneven medical revolution* (Rochester: University of Rochester Press, 2003); A.B. Davis, 'Life Insurance and the Physical Examination: A chapter in the rise of American medical technology,' *Bulletin of the History of Medicine*, 55 (1981), 392–406.

6 C. Timmermann, 'Clinical Research in Post-War Britain: The role of the Medical Research Council,' in C. Hannaway (ed.), *Biomedicine in the Twentieth Century: Practices, policies and politics* (forthcoming); J. Austoker and L. Bryder (eds), *Historical Perspectives on the Role of the MRC* (Oxford: Oxford University Press, 1989); C.C. Booth, 'From Art to Science: The story of clinical research,' in *A Physician Reflects: Herman Boerhaave and other Essays* (London: Wellcome Trust Centre for the History of Medicine at UCL, 2003), 79–101; C.C. Booth, 'Clinical Research,' in: W.F. Bynum and R. Porter (eds), *Companion Encyclopedia of the History of Medicine* (London & New York: Routledge, 1993), 205–29.

7 Cf. A.M. Brandt and M. Gardner, 'The Golden Age of Medicine?' in R. Cooter and J. Pickstone (eds), *Companion to Medicine in the Twentieth Century* (London: Routledge, 2003), 21–37.

8 On interactions between science and industry, see J.P. Gaudillière and I. Löwy (eds), *The Invisible Industrialist: Manufactures and the production of scientific knowledge* (Houndmills: Macmillan, 1998).

9 N. Rasmussen, 'The Drug Industry and Clinical Research in Interwar America: Three types of physician collaborator,' *Bulletin of the History of Medicine*, 79 (2005), 50–80; idem, 'The Moral Economy of the Drug Company-Medical Scientist Collaboration in Interwar America,' *Social Studies of Science*, 34 (2004), 161–85.

10 Cf. V. Quirke, 'From Evidence to Market: Alfred Spinks's 1953 survey of fields for pharmacological research, and the origins of ICI's cardiovascular programme,' in V. Berridge and K. Loughlin (eds), *Producing Health: medicine, the market and the mass media in the twentieth century* (London: Routledge, 2005), 144–69.

11 G.H. Acheson, 'Tetraethylammonium, Ganglionic Blocking Agents, and the Development of Antihypertensive Therapy,' *Perspectives in Biology and Medicine*, 19 (1975), 136–48.

12 Acheson, 'Tetraethylammonium,' p. 146.

13 Acheson, 'Tetraethylammonium'.

14 W.D.M. Paton, 'Hexamethonium,' *British Journal of Clinical Pharmacology*, 13 (1982), 7–14. See also H.P. Rang and P. Walton, 'Sir William Drummond MacDonald Paton, CBE, 5 May 1917–17 October 1993,' *Biographical Memoirs of Fellows of the Royal Society*, 42 (1996), 290–314. Paton's laboratory notebooks can be consulted in the Wellcome Library for the History and the Public Understanding of Medicine, PP/WDP/C/1/2. On the NIMR, see Austoker and Bryder (eds), *Historical Perspectives*, pp. 35–57.

15 Timmermann, 'Clinical Research'.

16 Paton, 'Hexamethonium,' pp. 7–8. See also Paton Laboratory Notebooks, Wellcome Library, PP/WDP/C/1/2.

17 Beta-blockers are a good example. According to the original physiological theory explaining their action, they should not have lowered blood pressure. Cf. Timmermann, 'To Treat or not to Treat'; V. Quirke, 'Putting Theory into Practice: James Black, receptor theory and the development of the beta-blockers at ICI, 1958–1978,' *Medical History*, 50 (2006), 69–92.

18 Paton, 'Hexamethonium,' p. 8.

19 Paton noted later that 'Barlow and Ing have never at any time either looked for or been interested in the ganglionic actions of this series'. Paton to Green, 19 July 1950, UK National Archives (hereafter UK NA), FD1/1172.

20 R.B. Barlow and H.R. Ing, 'Curare-like Action of Polymethylene *bis*-Quaternary Ammonium Salts,' *Nature*, 161 (1948), 718; W.D.M. Paton and E. J. Zaimis, 'Curare-like Action of Polymethylene *bis*-Quaternary Ammonium Salts,' *Nature* 161 (1948), 718–9.

21 Laboratory notebooks, Wellcome Library, PP/WDP/C/1/2.

22 Paton, 'Hexamethonium,' p. 10.

23 *Ibid*.

24 *Ibid*., p. 11.

25 UK NA, FD1/1172.

26 'The Methonium Compounds' (typescript), Paton Papers, PP/WDP/C6, Wellcome Library.

27 G. Organe, W.D.M. Paton and E.J. Zaimis, 'Preliminary Trials of Bismethylammonium Decane and Pentane Diiodide (C10 and C5) in Man,' *Lancet*, 253 (1949), 21.

28 P. Arnold and M.L. Rosenheim, 'Effect of Pentamethonium Iodide on Normal and Hypertensive Persons,' *Lancet*, 254 (1949), 321–3.

29 *Ibid*.

30 *Ibid*.

31 A.R. Hunter, 'Hexamethonium Bromide,' *Lancet*, 255 (1950), 251–2, p. 252.

32 Conference on Clinical Tests of Methonium Drugs, 22 June 1950, Minutes of the Meeting, UK NA, FD1/1172.

33 R. Turner, '"Medical Sympathectomy" in Hypertension: A clinical study of methonium compounds,' *Lancet*, 256 (1950), 353–8. See also M.L. Rosenheim, 'Lability of Blood Pressure,' *Lectures on the Scientific Basis of Medicine, 1951–52*, 1 (1953), 96–115. On the rice diet, see 'Diet in Hypertension [editorial],' *Lancet*, 255 (1950), 549–60; 'Rice Diet in Hypertension [editorial],' *Lancet*, 256 (1950), 529–30. The MRC organized a study of the effects of the rice diet: D.R. Cameron, D.M. Dunlop, R. Platt, M.L. Rosenheim and E.P. Sharpey-Schaffer, 'The Rice Diet in the Treatment of Hypertension: A report to the Medical Research Council,' *Lancet*, 256 (1950), 509–13.

34 See R. Platt and S.W. Stanbury, 'Sympathectomy in Hypertension,' *Lancet*, 255 (1950), 659–9; 'Sympathectomy in Hypertension [editorial],' *Lancet*, 255 (1950), 768–9.

35 S. Saville, 'Pentamethonium in Hypertension,' *Lancet*, 256 (1950), 358–60.

36 A. Campbell and E. Robertson, 'Treatment of Severe Hypertension with Hexamethonium Bromide,' *British Medical Journal*, 256 (1950), 804–6.

37 R. Turner, '"Medical Sympathectomy" in Hypertension'.

38 Minutes of a meeting on clinical tests of methonium drugs held on 19 July 1950, UK NA, FD1/1172.

39 A.E. Doyle, 'Sir Horace Smirk: Pioneer in drug treatment of hypertension,' *Hypertension*, 17 (1991), 247–50, p. 249. Doyle joined Smirk's department in 1952.
40 Memo, probably by Green, 27 Sept 1950, UK NA, FD1/1172.
41 Green to Paton, 4 October 1950, UK NA, FD1/1172.
42 Rasmussen, 'The Drug Industry and Clinical Research'; idem, 'The Moral Economy of the Drug Company-Medical Scientist Collaboration'.
43 Another therapeutic enthusiast was Edward Freis in the US, who later organized the large-scale treatment trials for hypertension with the Veterans Association Hospitals.
44 F.H. Smirk, 'Organisation of a Hypertensive Clinic, More Particularly For Patients On Methonium Treatment,' October 1951, typescript, UK NA, FD1/1172, p. 4.
45 F.H. Smirk, *High Arterial Pressure* (Oxford: Blackwell, 1957).
46 F.N. Fastier, 'Biography: Sir Horace Smirk: Professor Emeritus,' *New Zealand Medical Journal*, 67 (1968), 258–65, p. 258.
47 On clinical science in Britain, see Booth, 'Clinical Research' and idem, 'From Art to Science'.
48 F.H. Smirk and W.H. Hall, 'Inherited Hypertension in Rats,' *Nature*, 182 (1958), 727–8.
49 Organe, Paton and Zaimis, 'Preliminary Trials'; P. Arnold and M.L. Rosenheim, 'Effect of Pentamethonium Iodide on Normal and Hypertensive Persons,' *Lancet*, 254 (1949), 321–3.
50 E.A. Murphy, 'Treatment of Hypertension with Hexamethonium Bromide,' *Lancet*, 258 (1951), 899–901.
51 Doyle, 'Sir Horace Smirk', pp. 249–50.
52 F.H. Smirk, 'Organisation of a Hypertensive Clinic,' p. 5.
53 *Ibid.*, p. 4.
54 F.H. Smirk, 'Instructions for Patients on C6 Injections,' October 1951, typescript, UK NA, FD1/1172. On compliance, see J. Greene, 'Therapeutic Infidelities: "Noncompliance" enters the medical literature, 1955–1975,' *Social History of Medicine*, 17 (2004), 327–43.
55 Smirk, 'Instructions for Patients on C6 Injections,' p. 1.
56 Smirk, 'Organisation of a Hypertensive Clinic,' p. 7.
57 Green to Paton, 25 April 1952, UK NA, FD1/1172.
58 Smirk, 'Organisation of a Hypertensive Clinic,' p. 9.
59 *Ibid.*
60 *Ibid.*
61 *Ibid.*
62 Green to Paton, 25 April 1952, UK NA, FD1/1172.
63 'Methonium and Hypertension [editorial],' *Lancet*, 257 (1951), 395–6, p. 395.
64 E.D. Freis, 'Methonium Compounds in Hypertension [letter],' *Lancet*, 257 (1951), 909.
65 Paton to Green, 28 April 1952, UK NA, FD1/1172.
66 F.H. Smirk, 'Hypotensive Actions of Hexamethonium Bromide and some of its Homologues: Their use in high blood-pressure,' *Lancet*, 260 (1952), 1002–5.
67 See, for example, K.H. Beyer, 'Chlorothiazide,' *British Journal of Clinical Pharmacology*, 13 (1982), 15–24. See also Quirke, 'From Evidence to Market'.

10
Greenhouses and Body Suits: The Challenge to Knowledge in Early Hip Replacement Surgery 1960–1982

Julie Anderson

Introduction

Throughout the 1960s, increasingly complex surgical procedures were developed in orthopaedics, neurosurgery, cardiac, thoracic and eye surgery.[1] As these surgical processes were introduced, complications became evident which challenged the usual routines for anaesthesia or for infection control. This paper discusses one of the earliest of these new specialist types of surgery – hip replacement. It focuses on the peculiar problems of infection, the measures devised to limit the loss of operations, the debates around 'clean air,' and the consequences for other kinds of specialist surgery.

Hip replacement had been practiced very occasionally since the late-nineteenth century. However, it was not performed on any large scale until the early-1960s, when a series of innovations ensured that total hip replacement became common, replacing 'hemi-arthroplasties' (where the femoral head was replaced, or a cup placed over it) or hip fixation operations which sacrificed mobility to reduce the severe pain for arthritic joints. Despite the relative success of hip replacement surgery as a mechanical procedure, complications could compromise the results. One major problem was the rate of infection, a consequence of the long period of wound exposure, the implantation of prostheses, and the vigorous manipulation of both bone and muscle by the surgeon. Infection deep inside the tissue was hard to treat. A successful hip replacement could become a complete failure, as deep infection required the removal of the new hip joint and the performance of a salvage procedure, where the patient's mobility was often severely limited.

Research into the specific challenges of infection in hip replacement surgery was first conducted by John Charnley at Wrightington Hospital in Wigan near Manchester, and as I will demonstrate in this paper, his work showed that the quality of the air in the theatre played a pivotal role. Airborne pathogens which Lister had considered critical in the nineteenth century and which were subsequently downgraded in importance, were reexamined by Charnley who developed new practices and equipment designed to lower the risk from this source.

Surgical infection: historical background

The problem of infection was first addressed in the nineteenth century.[2] Joseph Lister was one of the first clinicians to apply Pasteur's theories on germs and contamination.[3] Lister's antiseptic practices, the use of phenol in the open wound, impregnated dressings and the air spray during surgery lowered his rate of infection, reducing death rates from 45.7 percent to 15 percent.[4] He also soaked his instruments and other equipment in carbolic, and while not scrubbing his hands, dipped them in a mixture of carbolic acid and antiseptic containing mercury. His early experiments with antiseptic practices on open fracture of the tibia, demonstrated the difficulties associated particularly with orthopaedics and infection.[5]

But the carbolic spray was soon abandoned by surgeons as it was difficult to set up and the concentration had to be carefully monitored lest it burn the skin.[6] In the twentieth century, concern over the causes of infection in surgery centred on the potentially infectious properties of the instruments, surgical materials, the operating theatre staff and patients' bodies. Aseptic practices after 1918 focused on protective clothing in the theatre, the sterilizing of instruments and the hygienic management of wounds. Routines such as hand washing and barrier methods to prevent contagion became established norms from the 1920s to the 1940s. The simple hand wash evolved into the more complex procedure of 'scrubbing up'. Gloves were worn as added protection against infection, first for the surgeon's benefit and then for the patient as well. Concern focused on how glove puncture might allow *Staphylococcus aureus* to leak onto the operating site.[7]

Hands were not the only vectors of infection, as tests demonstrated that contamination from the surgical team's bodies added to the risk. Droplet infection from theatre staff became a cause of concern, but masks were not routinely used; for example in 1910 it was suggested that surgeons should wear one only if they had a cold.[8] By the middle of the 1920s, *Streptococci* from nose and throat swabs was recognized as a potential source of infection; in one study published in 1935, the use of masks by all theatre personnel had significantly reduced streptococcal infections.[9] But many surgeons were complacent about their practices; for example, one surgeon in 1935 bemoaned the lack of 'infection consciousness'.[10]

The risk of transmission of infection to the patient by the staff was extended from the theatre to practices in the wards and dressing rooms. A report by the Medical Research Council's (MRC) War Wounds Committee in 1941 advised all personnel of new dressing techniques, to limit the exposure to bodily contamination, including hand washing between cases, and wearing a mask while changing dressings.[11] The focus of surgical transmission in the 1930s and 1940s was mainly on bodily contamination; the 'air' that concerned medical practitioners was that carrying bacteria from noses and throats. One article from 1941 stated, 'It seems probable that the risks of droplet infection from unguarded throats and of contact infection from unguarded fingers are greater than that of bacteria falling from the air.'[12] Nonetheless, there was still discussion about the role, and experiments were conducted to assess the role of the air in contamination.[13]

The development of antibiotics heralded a new dawn. With these 'magic bullets', it seemed that complications and deaths from infection would be almost eradicated. But from the beginning, cracks began to appear in antibiotics' armour of invincibility. Studies demonstrated that there was little improvement in rates of surgical infection from 1936.[14] Antibiotic-resistant strains of septic bacteria were causing concern as early as 1947, and awareness of this problem was heightened with several severe outbreaks of infection from *Staphylococcus aureus* in hospitals in the late-1950s.[15] One could not relay on the treatment of infections; infection-control remained critical.[16] Between 1957 and 1959, the Public Health Laboratory conducted a survey of surgical wound infection in British hospitals; rates were higher when the patient was older, the incision larger, the duration of the procedure was extended, and when drainage tubes were used.[17] The highest rates were for cholecystectomy at 21 percent. That the lowest were for orthopaedic surgery, 2 percent,[18] was because operative procedures at the time were largely non-invasive, involving the union of fractured bones.[19] By the middle of the twentieth century in Britain, mainly as a result of studies conducted by the MRC, recommendations on operating theatre design were made, though how infection was acquired remained controversial.

Operating theatre suites which had been built in the 1920s and 1930s provided their own unique problems. The extractor fans that installed in theatres to remove the heat drew in potentially contaminated air and steam produced by the in-suite sterilizers.[20] And there was relatively little new hospital building in the late-1950s and early-1960s because capital expenditure was limited and the National Health Service (NHS) concentrated on service delivery. It was difficult to build expensive new operating suites even for general surgery. But despite these fiscal restraints, there were changes to practices and a concentration on improving operating theatre design in order to prevent infection, with calls, in the 1960s, for standardization which will be addressed later.[21]

New suites built in the 1960s and 1970s were designed with plenum air-flow systems, where the operating theatre was at a higher pressure than the air outside it, so preventing the inward flow of bacteria laden air. These systems supplied heated humidified air, but they did not use filtration systems to remove contaminants.[22] These mechanical ventilation systems were generally the preferred method of reducing the effect of airborne bacteria in Britain, although air bacterial counts were not always good indicators of the likelihood of infection.[23] Other, so-called etiquette-related methods were also adopted, such as waiting five minutes to reduce air turbulence, ensuring that no staff entered or left the theatre, and preventing nurses from moving about by having all surgical materials ready and in easy reach. Other studies focused on factors such as the time of day of the operation. For instance, one study demonstrated that since operating theatres were used for a myriad of surgical procedures, particles built up over the day which meant that the potential for infection increased as the day wore on.[24]

In operations of a short duration or of limited exposure of tissues and organs, the fauna and flora of operating theatre air had not been of much consequence, but that changed with hip replacement surgery and led to the regulation and management of all aspects of the operating theatre's atmosphere and design. New specialist surgery in areas such as orthopaedics led to changes in operating theatre design, operating equipment and procedures. Fears about airborne transmission that had dominated the last decades of the nineteenth century resurfaced as a critical factor.

Other new types of orthopaedic surgery raised similar concerns. Open treatment of fractures by groups such as the Arbeitsgemeinschaft für Osteosynthesefragen (AO) who pioneered the plate and screw system, demonstrated the importance of surgical technique in the prevention of infection.[25] In this new type of surgery, developed in the 1950s and 1960s, the patient's body was also exposed for a long period of time, with vigorous activity on the part of the surgeon and the implantation of a relatively large foreign body.[26]

One of the biggest changes was the nature of the procedure. Instead of removing a malfunctioning or perhaps toxic body part, hip replacement was about the exchange of a faulty body part for a mechanical one. But unlike general surgery, any infection, which manifested itself, was located deep inside the tissue and was more difficult to resolve than a surface infection. A hip replacement could mean complete freedom of pain and resumption of function for a patient. If the replacement became infected, the result for the patient could be much worse than before the surgery was performed: 'the advent of infection can reduce the procedure from a singular success to a catastrophe.'[27]

John Charnley and clean air

Many of the early inventions and innovations in hip replacement surgery and the development of these new systems centred on John Charnley at the Wrightington Hospital near Wigan in Lancashire, England.[28] In his early cases, Charnley had an infection rate of 9 percent; he then began a programme to substantially reduce the rate. In 1961 he consulted Hugh Howorth, the grandson and owner of a pioneering filtration manufacturer in the north of England to construct a filtration system for the enclosure that he had already designed and built at Wrightington.

Using technology or practices from other types of industry was fairly common and hip replacements have benefited from the innovations in other industries. One of these was the clean air system. It was first developed in the late-nineteenth century, to clear the foggy air in cotton mills so as to keep the mills working for much of the year. In the twentieth century, the brewing industry also benefited from new filtration systems which kept air-borne contaminants from the fermentation process. Clean air enclosures were designed for other industries such as component manufacture. New types of ventilation were also being developed in other countries. In the United States this technology was introduced in the 1950s and 1960s particularly to prevent particle contamination in industry.[29]

In 1962, Wrightington Hospital was provided with its first clean air system operating theatre. The resultant design based its premise on laminar flow which is when the air flows in one direction so there is no cross current contamination. The design of these clean air systems had to be different from the conventional operating theatre as the contamination came from within the operating space and not external to it as in previous filtration designs. In order to prevent contamination, the laminar system was housed in an enclosure. While suitable for general surgery, plenum air or the pressured space, was demonstrated to be a hazard in implant surgery because it created air turbulence around the operation site.[30] This air turbulence caused bacteria to blow over the operation site and as the incision was about nine inches long there was a great deal of the body exposed to settling bacteria.

The first enclosure at Wrightington in 1962 was a small room measuring seven feet square and seven feet high. Air was ducted from outside. It was portable, and was set up on the days when Charnley performed surgery. Nicknamed 'The Greenhouse' for its diminutive dimensions and built by the engineer who worked at Wrightington Hospital in the laboratory, the prototype for the extraction and ventilation system was designed and donated by Howorth Air Systems. The air was cooled and the current strong with air changes of 100 times an hour. The enclosure's plate glass sides ensured that observers could see the procedure without coming into close contact with an open incision. Only three people were in the enclosure and a hatch at one end allowed the tray system of instruments to be passed to the surgeon. It was used for about two years and after it had been judged to be successful, Charnley applied for a grant from the MRC in 1963 to build a sterile operating enclosure at Wrightington citing his successes in reducing rates from 6.3 percent to 1.1 percent in the grant application.[31] In his application, Charnley appealed to the MRC's desire to be at the forefront of new technological developments and used the argument that he was also anxious to build the enclosure as he felt that the UK would be 'overtaken in this field' by others such as Maurice Muller in Berne, who had visited Wrightington and was apparently planning to incorporate a sterile enclosure in his new £1 million orthopaedic department.[32]

A sterile enclosure was installed at a cost of £3,199 which came from the successful application to the MRC.[33] This prototype endured many changes before another was permanently installed at Wrightington in 1966. The results were very good and Charnley claimed by using the system his infection rate had fallen to less than 1 percent.[34]

Throughout the 1970s the design of the Howorth Clean Air System changed. By 1973 the Charnley-Howorth air-system known as the

Down-Flow was larger, at 10 feet by 10 feet, and was also manufactured in polycarbonate which was easier to clean and to move, which worked well for the expanding export market. By 1976 the side walls were dispensed with by the development of the Ex-Flow, which was a zone as opposed to a room.[35] This Ex-Flow won a Design Council Award for Excellence in 1977. While the designs of the Charnley Howorth system changed between the early-1960s and the end of the 1970s, many of the underlying principles stayed the same.

While clean air systems were not installed in all hospitals in Britain and some orthopaedic surgeons continued to perform surgery in conventional theatres, a significant number of hospitals purchased the systems in the 1970s and 1980s. In the 1980s some of the 105 Howorth Air Systems installed included the Royal National Orthopaedic Hospital in Stanmore and RAF Ely. The Howorth systems were also exported to other countries. A total of 216 were installed in various countries, the majority in France (47) and the United States (38).[36] Howorth did not have a monopoly on the market for clean air enclosures, as other companies had also seen the opportunities for clean air systems and still others imitated the Howorth and other early designs. In the United States, the Laminar Flow Ultra Clean Room System manufactured by Agnew and Higgins in California and supplied through DePuy was developed in the early-1970s and were priced at approximately US$10,000 per unit.[37] Other orthopaedic manufacturers in the United States such as Howmedica supplied systems such as the Laminaire 700, and others developed their own systems including Richards. In Germany, Gellman were also developing clean air systems for orthopaedic surgery. As with other products, the Howorth systems did from association with Charnley, who was such an important name in the field of hip replacement surgery.

Other infectious sites

While the clean air system and isolation techniques practiced in Wrightington and other centres served to reduce the rates of contamination, there remained other sites for infection. The physical effort required of the orthopaedic surgeon in the 1960s was considerably higher than years later when instruments such as electric saws became more common. Surgeons had to use tools such as brace and bit which required a certain amount of strength. The effort of surgery caused the surgeon's body to give off about twice the amount of heat than when the body is at rest.[38]

Research had been conducted as early as 1948 into the potential infection of the patient by skin squames,[39] and these early results were confirmed by Charnley. Tests in the clean air enclosure at Wrightington demonstrated that the surgeon was an infectious source and the culprit was the epithelial cells from the surgeons' skin. Charnley's theory was that operations that required physical effort on the part of the surgeon tended to have a higher rate of infection than those that did not. He experimented in the clean air enclosure after surgery, taking samples off the surgeon's gown in the upper region of the body and the arms.[40] Research had demonstrated that the surgeon's whole clothed body as opposed to the exposed portions such as the face and the other common source of infection, the nasal-pharynx, were sources of infection. With that theory in mind, Charnley designed a new type of gown that surgeons could use in the operating theatre to ensure that the threat from infection via the skin route was minimized.

Charnley altered the design of the operating theatre gown at first adding a hood and a pull-on style.[41] These early gowns were unbearably hot as the surgeon was working hard in a tightly woven cotton gown which did not breathe in order that no skin cells would escape and fall into the wound. This problem of overheating was not overcome until 1970 when the use of suction, which extracted expired air out of the restrictive gowns, was perfected. The surgeon connected himself to a hose which removed the air and kept the surgeon cool. The pressure in the gown had to be sub-atmospheric so that the bellows action of the raising and lowering of the arms did not push skin cells into the air. While the body exhaust suits worked well in preventing infection, the ventile material was still hot and uncomfortable to wear. Woven cotton material, which had been the choice for most surgeons, was proven to be less reliable at preventing infection, partly because cotton was shown to shed more fibres than synthetic materials.[42] New woven materials designed in a similar way to the filters in the ventilation systems meant that particles found it difficult to pass through the fibres of these new synthetic gowns. The same design principles had been adopted with surgical masks where tests as early as 1935 had demonstrated that layers were one of the best design systems against the passing of bacteria.[43] At Wrightington, Charnley insisted on the one-piece ventile suit as the best way of preventing the spread of infection through skin cells from the surgeon to the patient.

Concerns about the body exhaust suits' efficacy were raised by surgeons however, who felt uncomfortable and who found the noise level of the extraction system made communicating difficult with their surgical team during the procedure. The shouting and mimed signals that Charnley used were not acceptable to all surgeons. Communication was required between an operating team who was unfamiliar with a procedure, so microphone systems were installed to solve this particular problem. Design changes were suggested, including a gown that tied at the back, and separate hood. Charnley was unhappy about these proposals, believing the patients' lowered risk of infection was more important than the surgeon's comfort. He wrote in the *British Medical Journal*, 'What I find alarming is the possibility that surgeons who do not understand the system may put pressure on the manufacturers and purely in the interests of their personal comfort may destroy the efficiency of the installation from the bacteriological point of view'.[44]

Not only was the design of the operating theatre space and clothing altered for this new type of surgery, questions were raised about the efficacy of design norms of other apparatus. Studies in Sweden and in the United States demonstrated that UV light did not have an effect on rates of infection, but other types of lighting could cause the air to heat up and cause turbulence. New designs including disinfecting UV were designed as the heat from lights in the theatre caused turbulence. Other methods to control infection in hip surgery included new operating techniques: the practice of double-gloving, wound closure including the abandonment of the t-incision in favour of a straight one, and stitching the subcutaneous fat layer.[45] In other centres surgeons used antibiotic prophylaxis and cement loaded with antibiotics. Charnley felt that anti-coagulant drugs increased infection so they were abandoned at Wrightington but used extensively elsewhere, particularly in the United States.

The MRC, the Department of Health and the NHS

As a result of the preoccupation with bacterial content in the air, coupled with the high rates of infections in general surgery, committees were established, reports written and recommendations made throughout the 1960s.[46] The high cost of infection and the increased number of 'bed-days' associated with it, pushed the cash-strapped NHS, research bodies and surgeons into action. In 1962 the MRC Subcommittee on Operating Theatre Hygiene recommended a zoning system where the operating theatre was surrounded by an outer protective area which was merely 'clean', whereas the theatre itself was sterile.[47] A disposal zone was added to minimize the risk of infection from soiled materials. The Royal College of Surgeons, who published their own report on the design of operating theatre suites in 1964, shared the MRC's concerns and made their own recommendations. Their report stressed the importance of climatic and environmental condition for both surgeons and the patient, and called for air-conditioning to be installed in operating theatres.[48] The Ministry of Health published its own report in 1967, endorsing general improvements, but playing down the need for air-conditioning in all operating theatres, no doubt due to the added cost of installing it in hospitals.[49] The new requirements were built into the hospital's design in the 1960s hospital building programme. The MRC Subcommittee, however, was still concerned with hygiene practices in hospitals and transmission of infection, and in 1968 another report was produced with recommendations for aseptic and hygiene systems in operating theatres. The stress on the operating environment signalled a shift away from the reliance on antibiotics, due mainly to the high number of antibiotic-resistant bacteria. In 1972, the MRC and the Department of Health formed a working party which produced a report on ventilation in operating theatres which led to a change in policy by the Department of Health to recommend air-conditioned operating theatres.[50]

Laminar flow systems were shown to reduce the percentage of infections at Wrightington, and many of the orthopaedic surgeons who trained there in the new two-part hip replacement, entered requests to their Boards for one of the systems that Charnley used with such good results. The system, at approximately £8,000 a unit, was relatively expensive to install. The body exhaust suits were an additional £2,500 which meant that the investment that a hospital was required to make was substantial and the cash-strapped NHS was often wary about investing in unproven technology. The alternative, antibiotics, were proven to cure most infections, and antibiotics did not involve capital investment in hospital space, although, as previously mentioned, the risk of bacterial resistance was of concern.[51] Other surgeons cast doubt on Charnley's results, suggesting that his lowered infection rates were a result of perfecting his technique. Charnley himself did not credit reduced infection rates wholly to the clean air room, and cited skin towels and suction drainage as also influencing his results.[52]

Various hospital boards had approached the Department of Health and Social Services 'for advice' on whether the systems were efficacious, which was why the Department of Health suggested a trial in 1972.[53] Charnley wrote to the MRC protesting this multi-centric investigation of Ultraclean air systems because he felt that all patients should be entitled to the best possible, procedure and to deny some patients would result in poor outcomes. The results of the 6,782 hip and 1,274 knee procedures from the MRC were summarized in a draft report: in a series of more than 8,000 operations for total joint replacement carried out in 19 hospitals between 1974 and 1979 the incidence of joint sepsis was approximately half for those operations done in an Ultraclean air system compared with that observed when a conventionally ventilated room was used. The wearing of whole-body exhaust suits in the Ultraclean air environment was associated with an even greater reduction, to less than one-quarter.[54]

So Charnley had been correct and his meticulous research had proven that for hip surgery, the environment was a significant factor in its successful outcome. Unlike general surgery, where airborne bacteria were a less significant source of infection, in prosthetic surgery the clean air system had a significant effect.

This new knowledge regarding infection and implant surgery was not accepted without challenge. One letter in the *British Medical Journal* in 1977 suggested that a decrease in the infection rate in the larger studies was due merely to the fact that the problem was being investigated and therefore staff were more aware and taking greater care to prevent infection.[55] According to some studies in the 1970s, laminar flow made little difference to infection rates in general surgery.[56] Other reports of controlled trials within the operating theatre environment were conducted; in one somewhat Listerian attempt, the air was sprayed with antibiotics to see if that would have any impact on infection rates.[57] In the face of protests that the environment in the theatre mattered less than surgical technique, laminar flow technology was adopted for different uses in the medical setting. The system was not just used in the operating theatre but was extended for use in other areas of medicine including burns units.[58] However, one question still remained, was it the environment or the discipline that the environment required that ensured that the infection rates remained so low?

Environmental concerns appeared to be less important in other countries. In the United States, other preventive measures were used in the battle against infection. While environmental considerations were important to some surgeons, others turned their back on the issue of airborne infection, preferring instead other methods to control the problem of infection in the hip. The infection, deep in the joint was very difficult for antibiotics to clear once it had taken hold, so antibiotic prophylaxis was used to curb infection.[59] In one study in the late-1970s, infection rates in hip surgery reduced from 25.7 percent to 1.2 percent using antibiotic prophylaxis.[60] Even though it had been demonstrated that laminar flow in joint replacement surgery had significantly reduced the number of infections, a report from the Public Health Laboratory published in the *American Journal of Medicine* in 1981 cast doubt that laminar flow would be any more effective than prophylactic antibiotics, or antibiotics introduced into cement.[61] Antibiotics became routinely used prophylactically and by 1973, a commentator noted that there was a 'definite trend' toward the use of antibiotics in hip replacement surgery in the United States.[62] In Britain, antibiotic prophylaxis gained creditability, but its use was adopted more slowly than in the United States, the importance of mechanical ventilation eclipsing other methods of infection control. Possibly one of the reasons why prophylaxis was taken up less readily in the UK was that Charnley strongly resisted the wholesale use of prophylactic antibiotics until 1982, although he did insert antibiotic powder (0.5 gram each penicillin and streptomycin) into the cavity of the replacement joint from 1965.[63] The question of resistant strains of bacteria was also a reason cited for a disinclination to routinely adopt prophylactic antibiotics. 'Antibiotics involve no installation costs, but when used prophylactically for each prosthetic operation they involve a continuing and cumulative expenditure and a potential, if small, hazard of emergent bacterial resistance'.[64] Despite resistance to the use of antibiotics prophylactically, they began to be used more regularly as an additional method of prevention, in conjunction with clean air systems. Eventually, Charnley viewed the use of antibiotics in surgery as a positive step in further reducing rates of infection. Charnley said in his address to the Hip Society in the United States that as infection was such a serious complication he believed that the combination of the two, clean-air and prophylactic antibiotics would hopefully lead to an elimination of infection in hip replacement surgery.[65]

Conclusion

The alterations researched and implemented by Charnley at Wrightington did not radically change all operating theatres or practices. The systems, designs and modalities that had served for general surgery were altered only incrementally, and clean air systems were not adopted for all operating theatres. New spaces and procedures were designed, although these were slow in development and implementation, partly due to the necessity of controlling costs in the NHS.

But the research conducted in trials in the 1960s demonstrated that the environment had more of an effect on some types of surgery than others, and also brought a realization that the tacit and established understanding surrounding the prevention of infection in general surgery would not suffice in the case of more complex surgical procedures such as joint replacement surgery. Surgeons and researchers like Charnley were striving for reduced rates of infection, as its results were so devastating, and in conjunction with the MRC-conducted trials and tests in order to lower risks for patients. More broadly, hip surgery identified new problems associated with bacterial control, increasing awareness of the effects of infection, the implication for the patient, and the cost to the NHS of increased hospital stays. New surgical practices, such as joint replacement surgery, called for the development of different techniques for infection control, mainly due to the invasive nature of the surgery. While antibiotic use, both as treatment and prophylactically, reduced infection rates in the operating theatre, the use of laminar flow and controls to limit the patient's exposure added to the surgeon's arsenal.

The new joint surgery changed the medical profession's attitude to infection, and coupled with the fear of antibiotic resistant strains of bacteria, changed practices both inside and outside the operating theatre space. As one researcher noted in 1973, 'It is reassuring to note that this procedure, carried out under the most vigilant antiseptic precautions, presents no greater risk of infection than other major orthopaedic procedures.'[66] Further, these developments in clean air surgery had implications for other complex procedures, including surgery on the heart, eye and brain. Research into the effects of clean air systems in these other fields of surgery demonstrated that for long procedures, such as open heart surgery, using clean air theatres reduced infection.[67] Over a period of time, studies continued to demonstrate that clean air impacted on the rates of surgical infection in orthopaedic surgery.[68] But as we have seen, the efficacy of clean air systems for operating theatres was challenged, and continued to be so after its adoption for orthopaedic surgery and other more complex procedures. Debates still surround the use of clean air systems and protective surgical garments.[69]

Acknowledgments

I would like to thank the Innovative Health Technologies programme of the Economic and Social Research Council for providing the funding for this research. I would also like to thank Professor John Pickstone and Professor Michael Worboys for their assistance with the preparation of this paper.

Notes

1   R. Faircliff, 'The Objectives in Planning Operating Theatre Suites,' in I.D.A. Johnson and A.R. Hunter (eds), *The Design and Utilisation of Operating Theatres* (London: Edward Arnold, 1984), 1–21, p. 6.

2   For a definitive volume on germ theories of disease and early infection control, see M. Worboys, *Spreading Germs: Disease theories and medical practice in Britain 1865–1900* (Cambridge: Cambridge University Press, 2000).

3   See R.B. Fisher, *Joseph Lister 1827–1912* (London: Macdonald and Jane's Publishers, 1977); and F.F. Cartwright, *Joseph Lister, the Man who Made Surgery Safe* (London: Weidenfield and Nicolson, 1963).

4   J. Lister, 'On the Antiseptic Principle in the Practice of Surgery,' *British Medical Journal*, 2 (1867), 246–8.

5   J. Lister, 'On a New Method of Treating Compound Fractures, Abscesses etc., With Observations on the Conditions of Suppuration,' *Lancet*, 89 (1867), 326–9.

6   Worboys, *Spreading Germs*, p. 95.

7   In one study published in 1939, the frequency of glove-puncture was found to be as high as 24 percent. E.A. Devenish and A.A. Miles, 'Control of *Staphylococcus Aureus* in an Operating Theatre', *Lancet*, 233 (1939), 1088–94, p. 1094.

8   W. Rose and A. Carless, *A Manual of Surgery*, 7th edition (London: Balliere and Co. 1920), p. 268.

9   F.L. Meleney, 'Infection in Clean Operative Wounds: A Nine Year Study,' *Surgery, Gynaecology and Obstetrics*, 60 (1935), 264–76, p. 266.

10 Meleney, *Surgery of Gynaecology and Obstetrics*, p. 264.

11 *Medical Research Council War Memorandum*, No. 6, 1941. See also R.E.O. Williams *et al*, 'The Control of Hospital Infection of Wounds,' *The British Journal of Surgery*, 32:127 (1944–45), 425–31. See also, A. Miles *et al*, 'Hospital Infection of War Wounds,' *British Medical Journal*, 1 (1940), 855–9.

12 E.T.C. Spooner, 'Observations on Hospital Infection in a Plastic Surgery Ward', *Journal of Hygiene*, 38 (1941), 320–9, p. 328.

13 D. Hart, 'Sterilisation of the Air in the Operating Room by Bacterial Radiant Energy', *Surgery*, 1 (1937), 770–1. See also T.B. Rice *et al*, 'The Bacterial Content of Air in the Operating Rooms,' *Surgery, Gynaecology and Obstetrics*, 73 (1941), 181–92.

14 A.I.L. Maitland, 'Postoperative Infection,' *British Journal of Surgery*, 52:12 (1965), 931–40, p. 938.

15 For a detailed discussion, see Robert Bud's paper in this volume.

16 See M. Barber, 'Staphylococcal Infection Due to Antibiotic-resistant Strains,' *British Medical Journal*, 1 (1947), 863–5.

17 Public Health Laboratory Service, 'Incidence of Surgical Wound Infection in England and Wales,' *Lancet*, 276 (1960), 659–63, p. 663.

18 Only one orthopaedic hospital was included in the survey and very few hip replacements had been performed at this stage. *Ibid.*

19 Orthopaedists often used surgery as a last resort. Surgical intervention on the tuberculous bones and joints of children in the 1930s had not yielded

good results. See L. Klenerman, *The Evolution of Orthopaedic Surgery* (London: The Royal Society of Medicine Press, 2002) and R. Cooter, *Surgery and Society in Peace and War: Orthopaedics and the organisation of modern medicine 1880–1948* (Hampshire: Macmillan, 1993).

20 O.M. Lidwell, 'Bacteriological Considerations,' in I.D.A. Johnston and A.R. Hunter (eds), *The Design and Utilisation of Operating Theatres* (London: Edward Arnold, 1984), 22–38, p. 30.

21 'New Surgery Brings Infection Risk', *The Times* (December 9, 1967).

22 C.C. Scott, 'Laminar/Linear Flow System of Ventilation,' *Lancet*, 295 (1970), 989–93, p. 989.

23 G.A.J. Ayliffe and M.P. English, *Hospital Infection: From miasmas to MRSA* (Cambridge: Cambridge University Press, 2003), p. 176.

24 The tendency to put 'dirty' cases on last, increasing contamination as the day wore on and fatigue on behalf of the operating team were given as potential reasons for this increase. J.S.S. Stewart and D.M. Douglas, 'Wound Sepsis and Operating-List Order', *Lancet*, 280 (1962), 1065–6.

25 The AO surgeons presented research which suggested that the experience of the surgeon was also a factor in rates of infection. T. Schlich, *Surgery, Science and Industry* (Basingstoke: Palgrave, 2002), p. 122.

26 Arguments were put forward that Laminar/Linear flow or high-exchange-rate operating theatres would not lead to a reduction in sepsis rates in general surgery. See D. Shaw, C.M. Doig, D. Douglas, 'Is Airborne Infection in Operating Theatres an Important Cause of Wound Infection in General Surgery?' *Lancet*, 301 (1973), 17–19.

27 I.D. Learmonth, 'Prevention of Infection in the 1990s,' *Orthopaedic Clinics of North America*, 24:4, (1993), 735–41, p. 735.

28 See J. Anderson, 'Innovation and Locality: Hip Replacement in Manchester and the North West of England,' *Bulletin of the John Rylands University Library* (forthcoming 2006)

29 Scott, 'Laminar/Linear Flow System of Ventilation,' p. 990.

30 S. Selwyn, 'Airing Operating Theatres,' *British Medical Journal*, 327 (1986), 1544–5.

31 *Proposals for a Sterile Operating Enclosure at the Centre for Hip Surgery, Wrightington*, UK National Archives, 21/1/64, FD 23/616.

32 *Proposals for a Sterile Operating Enclosure at the Centre for Hip Surgery, Wrightington*, UK National Archives, 21/1/64, FD 23/616.

33 Letter from John Charnley to the Medical Research Council, UK National Archives, 14/3/73, FD 23/616.

34 J. Charnley and N. Eftekhar, 'Post-operative Infection in Total Prosthetic Arthroplasty of the Hip,' *British Journal of Surgery*, 56:9 (1969), 641–9.

35 Advertising Leaflet Howorth Systems, Medical Collection, John Rylands Library, University of Manchester (undated: presumed 1976).

36 International Installation list Charnley-Howorth Air Systems, Medical Collection, John Rylands Library, University of Manchester (undated presumed 1989).

37 Advertisement Section, *Journal of Bone and Joint Surgery*, 53A:7 (October 1971), no page number.

38 Scott, 'Laminar/Linear Flow System of Ventilation,' p. 989.

39 J.P. Duguid and A.T. Wallace, 'Air Infection from Dust Liberated from Clothing,' *Lancet*, 252 (1948), 845–9, p. 845.

40   J. Charnley and N. Eftekhar, 'Penetration of Gown Material by Organisms from the Surgeon's Body,' *Lancet*, 293 (1969), 172–4, p. 173.
41   Charnley found that altering the style of the gown to the pull-on style and using the suction with the new style of gown made a significant difference in the bacterial content of the air. J. Charnley, 'Operating-Room Conditions,' Letter to the Editor, *Lancet*, 286 (1965), 907–8.
42   W. Whyte *et al*, 'The Reduction of Bacteria in the Operating Room Through the Use of Non-woven Clothing,' *British Journal of Surgery*, 65:7 (1978), 469–74, p. 469.
43   See C.J. Paine, 'The Aetiology of Puerperal infection With Special Reference to Droplet Infection,' *British Medical Journal*, 1 (1935), 243–6.
44   J. Charnley, 'Clean Air Operating Room Enclosures,' *British Medical Journal*, 1 (1974), 224.
45   J. Charnley, 'Wound Closure,' *Centre for Hip Surgery, Wrightington Hospital Internal Publication No. 33* (1971), 1–15.
46   No author, 'Hospital Infection,' *Lancet*, 276 (1960), 1235–6, p. 1235.
47   Operating Theatre Hygiene Subcommittee of the Committee on Control of Cross Infection, 'Design and Ventilation of Operating Room Suites for Control of Infection and Comfort', *Lancet*, 280 (1962), 945–51.
48   Royal College of Surgeons of England, 'Report on Design of Operating Theatre Suites,' *Annals of the Royal College of Surgeons of England*, 34 (1964), 217–88.
49   Ministry of Health, *Hospital Building Note No. 26: The Operating Department* (London: HMSO, 1967).
50   Medical Research Council and Department of Health, *Report by Joint Working Party on Ventilation in Operation Suites* (London: MRC and DHHS, 1972).
51   'Editorial: Airborne Infection in Surgery,' *Journal of Hospital Infection*, 3 (1982), 215–16, p. 215.
52   J. Charnley, 'A Sterile Operating Theatre Enclosure,' *British Journal of Surgery*, 51 (1964), 195–202, p. 202.
53   Minutes from the Ad-hoc Committee for Ultra-clean Air Systems in Operating Theatres, Collection of Mr K. Hardinge, 24/10/72.
54   Medical Research Council Trial Final Report, *Investigation of the Effect of Ultra-clean Air in Operating Rooms on Surgical Sepsis*, (London: MRC, 1979).
55   This is called the Hawthorn Effect. 'Letters', *British Medical Journal*, 2 (1977), 1244.
56   D. Shaw, C.M. Doig and D. Douglas, 'Is Airborne Infection in Operating Theatres an Important Cause of Wound Infection in General Surgery?' *Lancet*, 301 (1973), 17–20, p. 19.
57   The spray made no difference to rates of infection in general surgery. D.W Jackson, A.V. Pollock and D.S. Tindal, 'The Effect of an Antibiotic Spray in the Prevention of Wound Infection: Controlled trial,' *British Journal of Surgery*, 58:8 (1971), 565–6, p. 565.
58   R.H. Demling *et al*, 'The Use of a Laminar Airflow Isolation System for the Treatment of Major Burns', *American Journal of Surgery*, 136 (1978), 375–8.
59   See R.L. Nichols, 'Use of Prophylactic Antibiotics in Surgical Practice', *The American Journal of Medicine*, 70 (1981), 686–72.

60 New methods of Laminar flow ventilation were used which the authors admitted might have had an effect on their results. E. Simchen *et al*, 'The Successful Use of Antibiotic Prophylaxis in Selected High-risk Surgical Patients Under Non-trial, Everyday Conditions,' *Journal of Hospital Infection*, 1 (1980), 211–20, p. 211.

61 O.Lidwell, 'Airborne Bacteria and Surgical Infection,' *American Journal of Medicine*, 70 (1981), 693–7, p. 696.

62 F.E. Stinchfield, 'Editorial Comment: Total Hip Replacement,' *Clinical Orthopaedics and Related Research*, 95 (1973), 3.

63 J. Charnley and N. Eftekhar, 'Post-operative Infection in Total Prosthetic Replacement Arthroplasty of the Hip Joint,' *Centre for Hip Surgery, Wrightington, Internal Publication*, 17 (1968), 1–17.

64 'Airborne Infection in Surgery', *Journal of Hospital Infection*, 3, (1982), p. 216.

65 W. Waugh, *John Charnley: The man and the hip* (London: Springer-Verlag, 1990), 153–67.

66 Stinchfield, 'Editorial Comment: Total Hip Replacement,' p. 3.

67 See R. Lorenz, 'Air-conditioning Systems,' *Acta Neurchirurgica*, 55:1–2 (1980), 49–51; G. Soots *et al*, 'Air-borne Contamination Hazard in Open Heart Surgery: Efficiency of HEPA air filtration and laminar flow,' *Journal of Cardiovascular Surgery*, 23:2 (1982), 155–62; R.E. Clark, 'Infection control in cardiac surgery,' *Surgery*, 79:1 (1976) 89–96.

68 See M.A. Ritter *et al*, 'The Surgeons' Garb,' *Clinical Orthopaedics and Related Research*, 153 (1980) 204–9; F.H. Howorth, 'Prevention of airborne infections in operating rooms,' *Hospital Engineering*, 40:8 (1986) 17–23; P.E. Gosden, 'Importance of Air Quality and Related Factors in the Prevention of Infection in Orthopaedic Implant Surgery,' *The Journal of Hospital Infection*, 39:3 (1998), 173–80.

69 See S. Dharan and D. Pittet, 'Environmental Controls in Operating Theatres,' *Journal of Hospital Infection*, 51:2 (2002), 79–84; K. Verkkala *et al*, 'The Conventionally Ventilated Operating Theatre and Air Contamination Control During Cardiac Surgery: Bacteriological and particulate matter control garment options for low level contamination,' *European Journal of Cardio-thoracic Surgery*, 14:2 (1998), 206–10; D.C. Herman, 'Safety of a Clean Air Storage Hood for Ophthalmic Instruments in the Operating Room,' *American Journal of Ophthalmology*, 119:3 (1995), 350–4.

Part III

Expectations, Outcomes and Endpoints

11
From Epidemic to Scandal: the Politicization of Antibiotic Resistance, 1957–1969

Robert Bud

Introduction

The development of antibiotics in the 1940s has stood as one of science's greatest achievements in the control of nature for the good of humanity. Equally, the persistent ability of bacteria to develop resistance to antibiotics has been a standing reminder of the limits of such power. By the beginning of the twenty-first century, the general public had been made widely aware of the limitations of technical solutions to bacterial resistance and of the need to modify our behavior in the light of the threats of such bacteria as MRSA (Methicillin Resistant *Staphylococcus aureus)*. Their management even became an urgent issue in the 2005 British general election.[1] The threat itself, however, was not new. Resistant bacteria had been spreading, and their threat had worried public health experts since the beginning of the antibiotic age during World War II. Nor was the political interest unprecedented. While in the 1950s the general public in western countries had shown little interest in what was already a large problem, political and social concern with antibiotic resistance issues had grown radically in the late-1960s.

The changing climate of the 1960s was only partly the result of changing bacterial conditions. Much more important, I shall argue, were the changes in attitudes to the power of technology to transform nature. Particularly in the UK, by the late-1960s, journalists and broadcast media were drawing upon anxieties about diverse technologies to reinterpret the threat from resistant bacteria. The consequence was a major investigation of the application of antibiotics to farm animals, the Swann report, and subsequent restrictions on the agricultural use of these drugs, imposed in the hope that these would maintain the power of medicine to heal humans.

In this paper I shall contrast the experience of antibiotic resistance in the 1950s, which evoked little public response though it was major in scale, with the powerful public response to an epidemiologically moderate episode in 1967.[2] A newly sensitized media and political system transformed the *E. coli* outbreak in a small group of hospitals in the North-East of England into a political storm that led directly to the Swann report and the subsequent control of antibiotic use in animals.

Staphylococcus aureus

The 1950s' worldwide epidemic of an antibiotic-resistant strain of the bacterium, *Staphylococcus aureus* was experienced as major crisis by public health workers. The US Surgeon General told an emergency meeting held in October 1958 that 'every man and woman here knows that the stakes in this national problem are truly awful.'[3] The one bacterium was posing three different challenges: a nursery epidemic affecting newborn children, staphylococcal pneumonia particularly as a consequence of the influenza pandemic of 1957/58 and hospital-acquired infections following on surgery. On the other hand this epidemic seemed to create but little anxiety at the time to a general public used to relying on successful technical solutions to even the most difficult problems. Thus, Barbara Ronsencrantz reviewing 1950s literature for the US Office of Technology Assessment in the 1990s would find no anxiety about resistance.[4]

For many doctors, the shock of the encounter with resistant bacteria was particularly great because the recent development of antibiotics had seemed to transform medicine. In his 1953 *Natural History of Disease*, Macfarlane Burnet proclaimed, 'One can think of the middle of the twentieth century as the end of one of the most important social revolutions in history – the virtual elimination of infectious disease as a significant factor in social life'.[5] The Johns Hopkins surgeon Walsh McDermott later recalled the feeling of crossing 'an historic watershed'. He recalled how 'One day we could not save lives, or hardly any lives; on the very next day we could do so across a wide spectrum of diseases.'[6] This sense of power was widely associated with a relaxation of discipline in preventing infection both among patients and among doctors.[7]

The public in many western countries too had felt an epochal change in the threat from infection. This was summarized by the findings of the French sociologists Pierret and Herzlich who reported early in the 1960s that their older informants felt that 'real diseases' were no longer an anxiety. Many illnesses such as pneumonia which had once required lengthy periods of hospitalization could now be treated quickly at home. Certainly the emergence of antibiotics was not the only explanatory factor for secular decline in mortality from infection, but the statistics did support a greater sense of public confidence. In England and Wales hospital admissions for infectious disease fell from 85,000 in 1955 to 56,000 in 1960.[8]

The greater use of antibiotics was associated too with a multiplication of hospital operations and procedures not themselves associated with infection but now made relatively safer. Many more treatments, from cancer chemotherapy to kidney dialysis were becoming available. Thus, in the US, there was a 40 percent increase in hospital admissions during the period 1946–1961. In the years from 1946 to 1960 the number of beds rose by 16 percent but the number of admissions increased by 40 percent.[9] In Britain where the National Health Service had put hospital treatment within the range of more people the number of beds increased by 85 percent between 1938 and 1960.[10] The threats of cross-infection created by this increased turnover were again managed by antibiotics.

Resistance

The concentration of vulnerable patients in hospitals was providing a hospitable environment for bacteria, particularly *Staphylococcus aureus*. Certain strains were able to produce an enzyme, penicillinase, which destroyed penicillin, and apparently developed resistance to a variety of different antibiotics introduced during the early-1950s. As early as 1946 Mary Barber at the Hammersmith Hospital in London, was therefore monitoring bacterial susceptibility to penicillin. Generally, at first, the 'Staph. Aureus' she tested were susceptible to penicillin. Within two years, however, susceptibility declined from seven out of eight cases to three out of eight.[11] The solution Mary Barber proposed was careful control of antibiotic use. Under her guidance Hammersmith Hospital would pioneer the introduction of an antibiotic control policy in 1957. Nonetheless lesions, pneumonia and infant diseases caused by resistant *Staphylococcus aureus* beset hospitals and killed patients. Newborn infants were the first victims of the newly resistant bacteria to have problems which raised significant alarm within the medical community across the world. In the era of the baby-boom it was considered good practice to take infants from their mothers' sides and to put them into communal nurseries.[12] These proved to be outstanding market places for the exchange of bacteria and breeding grounds for the most successful.

A hospital in South Wales found that one in six babies experienced conjunctivitis or skin sepsis.[13] The children could go on to infect the breasts of nursing mothers which could in turn present a risk to a child after discharge. The experience of a hospital in Seattle in Washington State was reported in detail. Whereas only one death was attributed to the bacterium in 1953 and 1954, the problem worsened the following year.[14] That autumn in five Seattle hospitals at least 24 deaths were accredited to *Staphylococcus aureus*.

Diagnosis of the most virulent of strains responsible was carried out first in Australia, by Phyllis Rountree.[15] Having been trained in England she was familiar with the techniques that had been developed there, using phage types to fingerprint the bacteria they lysed. Rountree showed that the bacterium affecting Australian infants was the strain lysed by phage 80 and was very similar to a strain Canadian investigators had shown was lysed by phage 81. This bacterium which was shown to be the common factor behind the infection of infants throughout the world, was therefore termed 80/81. It combined resistance to penicillin and other antibiotics with virulence, causing a variety of hard-to-cure infections.

Surgical wounds in adults recovering in hospital were also prone to infection. This did not worry many surgeons responsible for the individual patient. Even if, typically 5 percent to 9 percent of clean surgical wounds resulted in some sort of infection, patients could in general be treated with higher and higher doses of antibiotics.[16] Britain's Chief Medical Officer Sir George Godber told a 1963 conference of a distinguished colleague who had denied there was a problem at all. Bacteriologists however placed the blame for increasing hospital infection on decreasing care of the surgeons.[17] As late as 1971 a sardonic article published by *The Annals of the New York Academy of Sciences* explained a new-found awareness of hard-to-treat infections caught in hospital. Until recently, it had not been 'done' in polite society to mention such problems. One explanation for a recent change in attitude, the authors suggested, might be '"an honesty transfer factor" passed from the younger to the older generations by a still unknown mechanism'.[18] Alternatively, they proposed, perhaps this new found openness was merely expedient at a time when environmental disquiet made any other response look complacent.

The most influential critic of current antibiotic use was Maxwell Finland at the Massachusetts General Hospital in Boston. Himself one of the first physicians to administer penicillin when he had been part of the team coping with the 1942 Cocoanut Grove fire in Boston now he was worried that antibiotic use was failing patients. In a 1959 paper he discussed the concurrent processes of increasing sophistication of operations and the growing use of resistance-reducing cancer chemotherapy multiplying the number of patients susceptible to infection and the increasing virulence of penicillin-resistant strains of *Staphylococcus aureus*.[19] The result was that the death rate of patients from staphylococcal infections had actually increased in the antibiotic age compared to pre-war rates.

Long term issues became a short term crisis in 1957 as threats from the same staphylococcus caused pneumonia complications associated with the influenza pandemic. In Britain the asian flu infected 9 million people during 1957, and a half of all school-age pupils succumbed.[20] The illness caused to most patients by the influenza virus itself was disruptive but not serious. However a minority of sufferers succumbed to secondary bacterial infections. Among these was a pneumonia which could be quickly fatal. Over one quarter of those infected, often once they had arrived in hospital, died of this infection. Because the number of influenza cases was very high, the number of deaths seem shocking. In November 1957 fatalities from pneumonia and influenza were twice what might have been expected in a normal year.[21] That winter, in the UK alone, there were 16,000 deaths beyond the number that might have been expected.[22] It has been estimated that, worldwide, a million people died.[23]

Response

The public health workers and bacteriologists taking a lead in responding to the *Staphylococcus aureus* challenge were appalled by the changes in hospital practice which encouraged the spread of resistant organisms. Thus Robert Williams, Director of the Streptococcal, Staphylococcal and Air Hygiene Laboratory at London's Central Public Health Laboratory, who could be compared to a central point of passage for the public health workers, led a long series of meetings each one emphasizing the need for more attention to hygiene.[24] In January 1957 he hosted a discussion of hospital infections under the auspices of the Society of Hospital Pathologists and one finds the same group of people organizing symposia in the early-1960s.[25] Repeatedly they uttered the mantra of urging greater reliance on prevention and traditional disciplines and less on antibiotics.

Similarly the US Centers for Disease Control (CDC) held an emergency meeting in Atlanta in October 1958. The Surgeon General followed through his military title and urged that what 'we must have throughout the country, is front-line officers and troops in an all-out attack'.[26] The first scientific paper was given by Robert Williams whose leadership was acknowledged by the surgeon general. The themes were accordingly similar to the British, and the *Proceedings* ended with a poem by Harry Dowling harking back to a previous age of infection management, verse four of which makes a pun of the English meaning of *Staphylococcus aureus*, 'Attention paid to grandma's rules/of cleanliness, for all but fools,/will decimate the golden horde/So none remain to bed and board'.[27] 10,000 copies of these proceedings were supplied to hospitals across the United States.[28]

The public however showed little concern and was not minded to contest the experts' policy, despite the worsening medical problem. Interesting evidence was provided through the clinical and even public opinion surveys prepared in the US as the 1957 influenza pandemic advanced across the world from the Far East. Despite the numerous complications and many deaths, a US Public Health Service study found that the public considered the epidemic less severe than anticipated.[29] One reason for this resilient confidence may have been the steep long-term decline in mortality from infectious disease, which was so radical that fatalities in the autumn of 1957 were less than the rate observed in a much less extensive epidemic as recently as February 1953.[30] Typically, when the *New York Times* ran a report on 1957 CDC findings, the headline was striking, 'Hospitals Found in Germ Danger. Resistance to Antibiotics is Cited for World Wide Epidemic', but the article itself just reported the papers presented at the meeting and there was no indication of any action expected of the general public or the government.[31]

In Britain the press became alarmed when it was announced in January 1958 that several hospitals were being forced to close wards on account of infestation by resistant *Staphylococcus aureus*. The first sentence of the article under the headline 'They Carry the Killer' in *The Daily Herald* for 14 January 1958 might not have gone amiss almost half a century later, 'Killer germ, "X", the mystery staphylococcus which is causing maternity ward death in hospitals around Britain, is believed to be carried by the very people who are trying to cure it'.[32] However while the report was on the front page, the albeit alarming words were given much less prominence than the main headline, 'Sarah Churchill in Jail'. So while civil servants noted such headlines and the half column devoted by the *Times* the same day to staphylococcus-caused deaths of influenza patients, they did not feel forced to react.[33] In general, whereas, during the influenza epidemic, doctors' concerns were reported, as in the United States, there was no implication that either the public or the government needed to react to public concern. Even a debate over hospital cleanliness was almost entirely reserved to technical journals, and the occasional letter to *The Times* failed to ignite public interest.[34] Albeit severe, the problems were treated as technical under the control of the relevant medical authorities who were, in general, coping successfully.

Indeed, the expectation of successful technical management of the problem of the 1950s might have seemed to be fulfilled. Hospital practice was modified. Infection control officers were introduced to many institutions.[35] The use of antiseptic hexachlorophene lotions to wash newborn babies had a significant impact on the nursery epidemic. Moreover, while bacteria might be resistant to penicillin, there were already other back-up antibiotics such as Erythromycin which could frequently be resorted to successfully. It is true that infections caused by the most resistant staphylococcal strain were not easily treatable, however in 1960 the Beecham and Bristol companies brought out their pioneering semisynthetic penicillin trademarked as Celbenin in the UK and Staphcillin in the US and which came to be known better as methicillin.[36] This provided cures even for infections caused by 80/81 *Staphylococcus aureus* and reassurance of potential technical superiority, although a strain of the bacterium resistant even to this new wonder drug (Methicillin Resistant *Staphylococcus aureus* – MRSA) did emerge almost immediately.

In the background, for no known reason, the 80/81 bacterium just became less common. The panic experienced by professionals themselves over its impact therefore subsided. The public, across the world had, in general, not shown any high propensity to worry. Their experience of new drugs, the decline of infections, and trust in professionals seem to have been a reassuring balm. Above all, perhaps, neither newspapers, nor television had taken up the story as a challenge for political action. More typical of that era was the February 1958 launch by the BBC of its highly successful series, 'Your Life in Their Hands' which reassured patients and potential patients by showing the public real-life operations for the first time.[37]

Technology in question

Ironically the fading of the threat from *Staphylococcus aureus* strain 80/81 early in the 1960s coincided with the decline of respect for authority in general and for the uses by which science was being deployed in particular. One sees not a rejection of science itself, but instead a reproach of the 'establishment' and resort to the outsider. Describing the condition of the US at the time, Margot Henriksen has explained the moods of revolution and antiauthoritarianism in contemporary literature and film, as means of coming to terms with the age of the nuclear bomb and the potential of total destruction.[38] Novels such as Thomas Pynchon's *V* or Kurt Vonnegut's *Cat's Cradle* or movies such as Hitchcock's *the Birds* or *Day of the Triffids* were darkly questioning of the management of the current technological age.

If the nightmares expressed by Hollywood could be dismissed as one walked out of the movie theatre, real life was less reassuring. Suddenly nothing was safe. Martin Bauer's studies of the contents of newspapers have documented the 1960s change in reportage of science and technology issues towards risk-warning critiques rather than positive affirmations.[39] This shift in style was so radical that it was quite clear at the time. A 1963 article in the US trade journal *Chemical & Engineering News* portrayed the evolution of science journalism from the pre-war 'Gee Whiz Age', through the post-war 'Reportorial Age' to the newly critical era now beginning which the author called 'Interpretive Age'.[40] In the United States, the chemical industry came under particular attack. The value of the industry's products in 1957 was more than six times greater than it had been in 1929.[41] Yet, tales of scientific achievement were suddenly superseded by popular concerns over the environment. The Manufacturing Chemists Association reported the sudden explosion of US media interest in environmental matters in a 1962 study.[42] This told member companies that, in recent years, ordinary people had begun to worry about the quality of their air and their water and they were blaming the chemical industry for polluting them. The moral drawn was that more public relations expenditure was required. Within months of the report, the industry's problems got much worse. Rachel Carson warned in her book *Silent Spring* of the danger of DDT, once hailed as a savior of mankind.[43] Exactly at the same time as her work was becoming public, it transpired that Americans had only barely been protected from thalidomide. Across the world the dangers of this drug to children born to women who had taken it became a scandal. In the same period, nuclear tests were revealed to release radioactive Strontium[90] and even the safety of nuclear power generation was questioned. In the UK, Ruth Harrison's denunciation of factory farming, *Animal Machines* published in 1964 told the public of such practices as the confinement of calves in veal crates.[44] This led not just to a widespread public outcry but also to the establishment of a committee by the Ministry of Agriculture, the Brambell Committee, and, subsequently, to a new law on Farm Animal Welfare. The technology of agriculture had become a matter of public debate and legislative action.

The Middlesbrough outbreak

Bacteria and antibiotics now also became the legitimate subjects of political action. The implications can be seen in the response to an antibiotic-resistance crisis which struck the North-East of England exactly one decade after the Autumn 1957 influenza pandemic.

In November 1967, hospitals in Middlesbrough noted that they were receiving a number of very young children suffering from gastro-enteritis caused by *E. coli* bacteria. More than 20 infants with the symptoms of gastro-enteritis were referred to just the West Lane Hospital. The infection spread within the hospital itself and infected another eight children. Again, a month later, 20 children were admitted with the same complaint. In compliance with normal procedure, the children were treated with the antibiotic neomycin. However their condition did not improve and instead declined markedly. Nor did they respond to a variety of other antibiotics: ampicillin (the broad spectrum penicillin), streptomycin, tetracycline, chloramphenicol, kanamycin or even sulfa drugs. That winter five babies died in West Lane Hospital.

The children had, in general, been weakened even before their illness. Four of them were very poorly, in any case, but one was a normal healthy baby whom the consultant paediatrician would have expected to survive. He had been admitted with pneumonia and a urinary tract infection on 23 November 1967. On 30 November he was diagnosed with enteritis apparently caught in the hospital and caused by the neomycin-resistant *E. coli*. He died on 22 December 1967. This local situation was worsened when children from an infected ward in the rapidly overloaded hospital were transferred to other local hospitals, taking the infection with them. As a result 11 children died in this epidemic, and of these most were severely handicapped.

Such epidemic spread of gram-negative infections was not unique to this hospital. It was a generic problem in developing countries but also across the developed world. In 1953/54, for instance, epidemic *E. coli* had spread up the East coast of the United States from South Carolina to New England and in 1961 about 5 percent of infants in the Chicago area were infected.[45] So, when on 19 December a press statement was made by the Middlesbrough General Hospital and the local Health Authority Secretary offered to meet local newspaper reporters, a routine local if sad story might have been expected.[46] Instead of reporters from two local papers, the meeting was attended however by numerous and aggressive journalists and by television.[47]

Two separate issues arose. One related to the admittance of children to a hospital known to be infected with the resistant *E. coli*. This was bad enough and led to an official investigation which highlighted a litany of small errors. Concern had also been compounded by the media, which, it was pointed out by the report, had not been very helpful.[48] The other, more profound, issue to emerge was the antibiotic resistance of the *E. coli*. When the outbreak was reported by the BBC's *Twenty-four hours* programme broadcast just before Christmas, national television introduced the public to a danger new to most people. Could the resistance to antibiotics in bacteria which kill children be caused by the feeding of antibiotics to animals?[49]

This programme brought to the attention of the public and politicians a campaign that had become increasingly vociferous over the previous five years. It had begun as a scientific speculation, aroused the interest of fellow scientists, it had won the support of science journalists, and now the link to the Middlesbrough deaths signalled a 'scandal' – a debate in the House of Commons provided the opportunity for Members of Parliament to express outrage and the press demanded action.[50]

Since the early-1950s, low doses of antibiotics had been administered to poultry and pigs to speed their growth. The use of such subtherapeutic doses was known to select for resistant organisms, however the difference between the pathogens affecting animals and people seemed to provide assurance that there were no dangers to human health. Between 1960 and 1962 a committee established by the British government under the chairmanship of the National Farmers Union's president, Lord Netherthorpe, reviewed this procedure and its final report generally provided support.

Transferable resistance

Just as the Netherthorpe report had appeared it had been undermined by studies conducted on the transmission of resistance between bacteria. While such studies had been conducted for some years in Japan, it was only in 1961 and 1962 that scientists in the West were made aware of them by Naomi Datta at the Hammersmith Hospital.[51] At the

Central Public Health Laboratory, it was shown how such processes explained the emergence of a strain of resistant *Salmonella Typhimurium*, a bacterium causing gastro-enteritis in humans, which was resistant to several antibiotics.[52]

The threat that haphazard antibiotic use in animals could select for genes that might then be transferred to bacteria threatening humans was raised. Such abuse could occur in the feeding of antibiotics prophylactically to calves sent to distant markets where they might mix with infected cattle, as well as in the use of subtherapeutic doses to enhance growth. Robert Williams would recall the suggestion that the *Salmonella typhimurium* outbreak was caused by a single cattle dealer who was an over-enthusiastic user of antibiotics and only ended on the dealer's untimely death.[53] Warnings coming out of the laboratory stimulated a recall of the scientific subcommittee of the Netherthorpe Committee which in turn recommended a full enquiry of agricultural uses of antibiotics.[54] This report by itself was soon buried in the crisis of the Foot and Mouth outbreak of 1967. On the other hand, by the Autumn of 1967 several journalists including Anthony Tucker, science editor of *The Manchester Guardian* and Bernard Dixon, soon to be editor of *New Scientist* had become worried.

Dixon himself published an article in *New Scientist* under the title 'Antibiotics on the Farm – major threat to human health' in October 1967. Dixon reported the 1965 findings that 590 humans had been infected by *Salmonella typhimurium* characteristic of animals and the danger of transmission of resistance to even more threatening organisms. Thus a subhead within the article read, prophetically, 'Diarrhoea in Babies'. Above all the article charged the authorities of inertia and of evading the implications of the scientific findings by their proposal to have yet another investigation. It begins in a way quite different from the respectful reportage of the treatments a decade earlier: 'So nothing is yet to be done to curb the reckless and dangerous exploitation of antibiotics in farming.'[55]

Dixon's article was still fresh when the Middlesbrough crisis struck. The children had been infected by two virulent strains of the *E. coli* bacterium commonly found in the gut and which could also be found in animals. Just five years earlier a small outbreak of typhoid in the Aberdeen area had been contained but if antibiotics could no longer be trusted, what then for the confidence of a modern country that such diseases were no longer to be feared?[56]

The government needed to act not least because ministers were themselves becoming concerned about antibiotic resistance. In January,

Shirley Williams, then junior minister of education, had been worried, a civil servant reported: '(possibly in the light of recent press reports that transferable resistance may have been a factor in the Teesside deaths of babies) about a press article which appeared recently recommending what looked like large-scale and indiscriminate use of antibiotics in the treatment of 'flu.'[57] A few days later it was noted 'Ministers are becoming increasingly vulnerable in this business and we ought quickly to settle our lines on Netherthorpe.'[58] As a result a new committee came under discussion.

The official report on the Middlesbrough crisis was completed in March 1968 but not circulated on the grounds that it mentioned individuals' names. This in turn raised suspicions. An angry adjournment debate was held in the House of Commons in April led by the Yorkshire Conservative MP Timothy Kitson who had long showed an interest in antibiotic resistance and whose constituency abutted Middlesbrough.[59] Quoting the *New Scientist* article at length, he assaulted the government both for downplaying the severity of the Middlesbrough outbreak and for allowing the issue of transferable antibiotic resistance to slip off the agenda. The Minister, Kenneth Robinson, could only apologize.

Unlike the speedy calming of a decade earlier, this time the press refused to lose interest. Even the tabloid *Daily Mirror* took up the cause of protecting antibiotics.[60] Civil servants in the department of health therefore revived the recommendations of the Netherthorpe committee, now two years old, and insisted on a formal enquiry into the use of antibiotics for animal feeding.[61] The outcome of this campaign was the committee chaired by Lord Swann which recommended that key antibiotics important to humans including penicillin and tetracycline should not be administered to animals solely to promote growth.[62] This report was considered highly controversial but further pressure from journalists meant that it did reach the statute book.[63] Indeed similar legislation was enacted across Europe.

Conclusion

The Swann Committee and its legislative consequences were therefore a direct result of the campaign waged on the back of the 1967 Middlesbrough outbreak. A decade earlier, numerous and even more deadly episodes of widespread infection affecting infants had had no such effect. Instead journalists had reported them as matters that only doctors could engage with. In part the change had occurred because the issue of resistance had widened from medicine whose practice was still generally beyond political debate to agriculture which was being widely discussed. However this translation had followed from the active intervention of journalists and politicians who had become used to a much more proactive engagement with science. The experiences of the early-1960s had been a turning point. By 1968, just 25 years after the introduction of penicillin, journalists and a more sceptical public had made antibiotic resistance an unavoidably political issue.

Acknowlegments

I am grateful to Bernard Dixon and to Professor Alan Linton for their advice and recollections of the origins of the Swann Report.

Notes

1 'Michael Howard: £10 million Technology to Fight Hospital Superbug,' Conservative Party Press Release, 26 April 2005. Also 'Action on Health,' Conservative Manifesto, chapter 3.
2 On the broader context see Robert Bud, *Penicillin: Triumph and tragedy* (Oxford: Oxford University Press, 2007).
3 L.E. Burney, 'Staphylococcal Disease: A national problem,' in *Proceedings of the National Conference on Hospital-Acquired Staphylococcal Disease, sponsored by US PHS and NAS* (Atlanta Georgia: US DHEW, CDC, October 1958), 3–10.
4 B. Rosencrantz, 'Coverage of Antibiotic Resistance in the Popular Literature 1950 to 1994,' in Office of Technology Assessment, *Impacts of Antibiotic-resistant Bacteria: 'Thanks to penicillin, he will come home!'* Publication OTA-H-6297 (Washington, D.C.: Office of Technology Assessment, 1995), Appendix.
5 Macfarlane Burnet, *Natural History of Disease*, 2nd edition (Cambridge: Cambridge University Press, 1953), p. ix.
6 W. McDermott, with D.E. Rogers, 'Social Ramifications of Control of Microbial Diseases,' *Johns Hopkins Medical Journal*, 151 (1982), 302–12.
7 See for instance, L.K. Martell, 'Maternity Care during the Post-World War II Baby Boom,' *Western Journal of Nursing Research*, 21 (1999), 387–404, p. 394. See also Bud, *Penicillin*.
8 G. Godber, 'Opening Address' in *Prevention of Hospital Infection: The personal factor. report of a conference held in London on 19 June 1963* (London: Royal Society of Health, 1963), 1–3.
9 S.E. Harris, *The Economics of American Medicine* (New York: Macmillan, 1964), p. 171.
10 D. Armstrong, 'Decline of the Hospital: Reconstructing institutional dangers,' *Sociology of Health and Illness*, 20 (1998), 445–57.
11 L.P. Garrod, 'Mary Barber 3 April 1911–11 September 1965,' *Journal of Pathology and Bacteriology*, 92 (1966), 603–10; M. Barber, 'Staphylococcal Infection Due to Antibiotic-resistant Strains,' *British Medical Journal*, 2 (1947), 863–5; idem, 'The Incidence of Penicillin Sensitive Variant Colonies in Penicillinase Producing Strains of *Staphylococcus aureus*,' *Journal of General Microbiology*, 3 (1949), 274, and a long sequence of further works.

12 E. Temkin, 'Rooming-In: Redesigning hospitals and motherhood in Cold War America,' *Bulletin of the History of Medicine*, 76 (2002), 271–98; W. Wheeler, 'Recollections of 40 Years of Hospitalization of Children,' *Pediatric Research*, 6 (1972), 840–2.

13 R.E.O. Williams, 'Investigations of Hospital-Acquired Staphylococcal Disease and its Control in Great Britain,' *Proceedings of the National Conference on Hospital-Acquired Staphylococcal Disease*, 11–29.

14 R.T. Ravenholt and G.D. La Veck, 'Staphylococcal Disease – An Obstetric, Pediatric and Community Problem,' *American Journal of Public Health*, 46 (1956), 1287–96.

15 On Rountree's work see P.M. Rountree, 'History of Staphylococcal Infection in Australia,' *Medical Journal of Australia*, 2 (1978), 543–6. Rountree described her career in a 1991 oral history conducted by Kerry Gordon: Victoria Barker (ed.), 'Phyllis Margaret Rountree,' University Interviews Project, University of New South Wales Archive.

16 Williams, 'Investigations of Hospital-Acquired Staphylococcal Disease'.

17 Godber, 'Opening Address'.

18 H.D. Isenberg and J.I. Berkman, 'The Role of Drug-Resistant and Drug-selected Bacteria in Nosocomial Disease,' *Annals of the New York Academy of Sciences*, 182 (1971), 52–8.

19 M. Finland, W.J. Jones Jr. and M.W. Barnes, 'Occurrence of Serious Bacterial Infections since Introduction of Antibacterial Agents,' *Journal of the American Medical Association*, 170 (1959), 2188–97.

20 Ministry of Health, *The Influenza Epidemic in England and Wales 1957–1958*, Reports on Public Health and Medical Subjects No. 100 (London: HMSO, 1960).

21 R.H. Drachman, G.M. Hochbaum and I.M. Rosenstock, 'A Seroepidemionic Study in Two Cities,' in I.M. Rosenstock, G.M. Hochbaum, H. Leventhal *et al* (eds), *The Impact of Asian Influenza on Community Life: A study in five cities*, Public Health Service Publication No. 766 (Washington D.C.: US Government Printing Office, 1960), pp. 26–51, see p. 35.

22 Ministry of Health, *The Influenza Epidemic in England and Wales 1957–1958*; for US figures see C.C. Dauer and R.E. Serfling, 'Mortality from Influenza, 1957–1958 and 1959–1960,' *American Review of Respiratory Diseases*, 83 (1961), 15–28.

23 See for instance 'Pandemic Flu. UK Health Departments' UK Influenza Pandemic Contingency Plan,' http://www.show.scot.nhs.uk/sehd/pan-demicflu/Documents/j6645e-13.htm (accessed February 2006); E.D. Kilbourne, 'Influenza Pandemics of the 20th Century,' *Emerging Infectious Disease* [serial on the Internet], January 2006. Available from http://www.cdc.gov/ncidod/EID/vol12no01/05-1254.htm (accessed January 2006).

24 For Robert Williams see his obituary in the *British Medical Journal*, 327 (30 August 2003), p. 506.

25 R.E.O. Williams and R.A. Shooter, *Hospital Coccal Infections* (London: Association of Hospital Pathologists, 1958) and 'Hospital Coccal Infections,' *Lancet*, 1 (1957), 92–3; Also see R.E.O. Williams, *Hospital Infection: Causes and Prevention* (London: Lloyd-Luke, 1960); the 1963 *Prevention of Hospital Infection. The Personal Factor*; and R.E.O. Williams and R.A. Shooter (eds), *Infection in Hospitals: Epidemiology and Control. A Symposium organized by the*

Council for International Organizations of Medical Sciences (Oxford: Blackwell, 1963).

26 Burney, 'Staphylococcal Disease,' p. 9.
27 H. Dowling, 'Summary of Conference,' in *Proceedings of the National Conference on Hospital-Acquired Staphylococcal Disease*, 175–8.
28 R.I. Wise, E.A. Ossman and D.R. Littlefield, 'Personal Reflections on Nosocomial Infections and the Development of Hospital Surveillance,' *Reviews of Infectious Diseases*, 11 (1989), 1005–19.
29 See Rosenstock, Hochbaum, Leventhal *et al*, *The Impact of Asian Influenza.*
30 Drachman, Hochbaum and Rosenstock, 'A Seroepidemiologic Study in Two Cities,' p. 35.
31 R.K. Plumb, 'Hospital Found in Germ Danger: Resistance to antibiotics is cited by surgeons for world wide epidemic,' *New York Times* (17 October 1957).
32 'They Carry The Killer,' *Daily Herald* (14 January 1958).
33 'Germs Resist Penicillin,' *The Times* (14 January 1958). See also 'Baby-Killer Germ Shuts More Wards,' *Sunday Express* (12 January 1958). These were reported in 'Today's News In Brief Tuesday, January 13th, 1958' and a similar memorandum the next day, in 'Standing Medical Advisory Committee, Sub-committee to Consider Antibiotic Resistant Staphylococcal Infection in Hospitals: report 1957–1960,' MH133/223, UK National Archives.
34 R. Granville-Mathers, 'Washing the Walls,' *The Times* (28 January 1959); F.A. Jones, 'Who Should Clean Hospital Walls,' *The Times* (2 February 1959); Mrs G.E. Hayes, 'Hospital Walls,' *The Times* (2 February 1959).
35 S. Mudd, 'Staphylococcic Infections in the Hospital and Community,' *Journal of the American Medical Association*, 166 (1956), 1177–8.
36 E.M. Tansey and L.M. Reynolds (eds), *Post-penicillin Antibiotics: From acceptance to resistance?* Wellcome Witnesses to Twentieth Century Medicine 6 (London: Wellcome Institute for the History of Medicine, 2000).
37 A. Karpf, *Doctoring the Media: The reporting of health and medicine* (London: Routledge, 1988).
38 M.A. Henriksen, *Dr. Strangelove's America: Society and culture in the Atomic Age* (Berkeley: University of California Press, 1997).
39 M. Bauer, '"Science in the Media" as a Cultural Indicator: Contextualizing surveys with media analysis,' in M. Deirkes and C. von Grote (eds), *Between Understanding and Trust: The public, science and technology* (Reading: Harwood, 2000), 157–78. See also D.P. Nord, 'Conventional Themes in Science Fiction,' *Journal of Popular Culture*, 13 (1979), 264–73.
40 L. Lessing, 'The Three Ages of Science Writing,' *Chemical & Engineering News*, 41 (6 May 1963), 88–92.
41 A. Thackray, J.L. Sturchio, P.T. Carroll and R. Bud, *Chemistry in America, 1876–1976* (Dordrecht: Reidel, 1984), p. 313.
42 Manufacturing Chemists Association Inc, 'A Report to the Board of Directors by the Public Relations Advisory Committee,' 14 February 1961, CMA 067883. This has been published on the web at www.chemicalindustryarchives.org (accessed 1 December 2004). For a recent general overview of the development of environmentalism in this period see H.K. Rothman, *The Greening of a Nation: Environmentalism in the United States since 1945* (New York: Harcourt Brace, 1998).

43   R. Carson, *Silent Spring* (New York: Houghton Mifflin, 1962). See L. Lears, *Rachel Carson: Witness for Nature* (New York: Henry Holt, 1997).

44   R. Harrison, *Animal Machines: The New Factory Farming Industry* (London: Vincent Stuart, 1964).

45   P.W. Ewald, 'Guarding Against the Most Dangerous Emerging Pathogens: Insights from evolutionary biology,' *Emerging Infectious Diseases*, 2 (1996), 245–56 summarizes the data.

46   Sidney Lightfoot details from 'The Men who Fought It,' *Evening Gazette* (Middlesborough, 7 March 1968).

47   'The Diary of a Tragedy,' *Evening Gazette* (7 March 1968).

48   C.A. Green and E.G. Brewis, 'An Enquiry into the Outbreak of *E. Coli* 0128 Gastro-Enteritis in Infants under One Year Old in Teesside', MH/60/788, UK National Archives, p. 18.

49   The programme was broadcast on 21 December 1967.

50   'Gastro-Enteritis on Teesside,' *Parliamentary Debates (Commons)*, 5th series, 762 (1967–68), 1619–1630, held on 11 April 1968.

51   Tansey and Reynolds (eds), *Post-penicillin Antibiotics*, 45; T. Watanabe and T. Fukosawa, 'Episome-mediated Transfer of Drug Resistance In Enterobacteriaceae I: Transfer of resistance factors by conjugation,' *Journal of Bacteriology*, 81(1961), 669–78; idem, 'Episome-mediated Transfer of Drug Resistance In Enterobacteriaceae II: Examination of resistance factors with acridine dyes,' *Journal of Bacteriology*, 81 (1961), 679–83; idem, 'Episome-mediated Transfer of Drug Resistance In Enterobacteriaceae III: Transduction of resistance factors,' *Journal of Bacteriology*, 81 (1961), 202–9, T. Watanabe, 'Infectious Drug Resistance,' *Scientific American*, 217 (1967), 19–28; T. Brock, *The Emergence of Bacterial Genetics* (Cold Spring Harbor, New York: Cold Spring Harbor Laboratory Press, 1990).

52   E.S. Anderson, 'Origin of Transferrable Drug-Resistance Factors in the Enterobacteriaceae,' *British Medical Journal*, 2 (1965), 1289–91.

53   R.E.O. Williams, *Microbiology for the Public Service: Evolution of the Public Health Laboratory Service 1939–1980* (London: Public Health Laboratory Service, 1985), p. 107.

54   I deal in greater length with the origins of the Swann Committee in *Penicillin: Triumph and Tragedy*. On the Netherthorpe Committee's second report see Agricultural Research Council and Medical Research Council, Joint Committee on Antibiotics in Animal Feeding. Second Report of the Scientific Subcommittee,' Appendix 1 to MRC 66/939 and ARC 22B/66 in MAF287/450, UK National Archives.

55   B. Dixon, 'Antibiotics on the Farm – Major Threat to Human Health,' *New Scientist* (5 October 1967), 33–5.

56   A.T. Roden to Dr Shaw, 18 January 1968, MH 160/788, UK National Archives.

57   G.B. Blaker to T.B. Wilkinson, 17 January 1968. FD7/ 899, UK National Archives.

58   T.S. Williamson to J. Hensley, 27 January 1968, FD7/899, UK National Archives.

59   'Gastro-Enteritis on Teesside.' Kitson's speech is reproduced on columns 1619–24.

60   'Action Sought on Antibiotics after Babies' Deaths,' *The Times* (14 April 1969); 'F and M at the Min of Ag,' *Daily Mirror* (29 April 1968).

61 Note of a Meeting with the Ministry of Agriculture Fisheries and Food',
    21 February 1968, FD7/900, UK National Archives.
62 M.M. Swann (Chair), *Report [of the] Joint Committee on the Use of Antibiotics
    in Animal Husbandry and Veterinary Medicine*, 1969–70, v Cmnd.4190
    (London: HMSO, 1969).
63 See B. Dixon, *What is Science For?* (London: Collins, 1973), pp. 137–40.

12
Cancer Clinical Trials and the Transfer of Medical Knowledge: Metrology, Contestation and Local Practice

Gerald Kutcher

Termites build their obscure galleries with a mixture of mud and their own droppings; scientists build their enlightened networks by giving the outside the same paper form as that of their instruments inside. In both cases the result is the same: they can travel very far without leaving home.[1]

Attempts to change the measure often encountered resistance, occasionally riot.[2]

Post-World War II American cancer researchers were struck by the ability of clinical trials to answer medical questions, to resolve controversy, and to advance medicine. These budding American researchers, who were schooled in large-scale cooperative medical research during the war, and who by the mid-1950s dominated clinical research in the US, believed that with the apparatus of a clinical trial, impartial observers using objective criteria would establish medical facts that would replace knowledge passed down by tradition. Antiquated attitudes would be swept away and replaced by a rational medicine based upon unbiased prospective randomized trials.

American clinical trials in the post-war era were run as a metrological practice: a coordinating centre promulgated protocol standards, disseminated them outward to the participating institutions, monitored protocol compliance to assure the maintenance of the standard, and then received and processed the clinical outcomes. Metrology is usually associated with the maintenance of standards for quantities like length, weight and time; in medicine metrology refers to the calibration of technologies from the thermometer to high-energy linear accelerators for radiation therapy. For scientific researchers and indeed for most historians of medicine, metrology, if it is considered at all, is seen as a rather narrow and specialized subject with little applicability. Yet, metrology has great importance in the functioning of all modern technoscience, a position that has been advanced by Bruno Latour, Simon Schaffer and others.[3] Moreover, historians such as Witold Kula and E.P. Thompson have shown the vital role that metrology played in the economic and political functioning of local communities and states. Standards for weights and measures, according to Kula, reside in diverse locations like the collective farming community, the medieval manor house, the royal household and the bureaucratic state[4] – and for our story, in standards located at consensus conferences of cancer specialists and among coteries of clinicians. The control of standards, to be sure, has wide-ranging economic, political and social implications, as we will see.

We should appreciate that metrology in medical practices differs from its more usual incarnation in the physical sciences where standards are disseminated in one of two ways. A standards laboratory (or other central authority) may maintain artefact standards and disseminate them by calibration, that is, by comparing them to secondary, tertiary and other lesser measuring systems. The meter, maintained by the government in post-revolutionary France, was disseminated through a number of standard length bars which were inter-compared with the primary standard and then embedded in the facades of buildings. Alternatively, intrinsic standards reside in procedures or experiments (defined by formal protocols), which if properly followed, result in the desired unit of measure. One unit for grain in the medieval period was defined by the following protocol: pour the grain 'from dropped-arm height striked and unpressed' into a container of prescribed dimensions with the container on level ground and without shaking.[5] In recent years, the unit for the meter, having previously resided in a material object, has been defined through an experimental protocol.[6] Since intrinsic standards are disseminated via a protocol or experiment, the auditing of documents outlining local procedures replaces inter-comparisons.

In spite of the highly controlled metrological standards, which were brought to bear on clinical trials, the knowledge claims that American medical researchers produced in the post-war period were almost always controversial. Attacks on the design, execution, ethics and results of clinical trials came from researchers, prominent physicians, clinicians and special interest groups. To counter the endless debates, by the mid-1970s the National Institutes of Health (NIH) put in place another centralized structure that mirrored clinical trial metrology, namely the consensus conference, whose purpose was to settle controversy and construct medical knowledge. In spite of the promise of their highly formalized and supposedly unbiased character, consensus conferences often could not reach closure. And when they did, the knowledge they constructed and the change in practices they demanded were contested at the local level by practicing physicians.

The difficulty of transferring knowledge claims and changing practices have been addressed by a number of historians of medicine. Some have argued that incommunicable knowledge is at the heart of the difficulty of persuading physicians with one set of interests and training to take up the knowledge claims of another group.[7] Others, using actor-network theory, have claimed that medical practice can indeed support, at one and the same time, diverse and consistent procedures.[8] While both these points of view are clearly present in medical practice, my emphasis here will be on the difficulty of reaching consensus. I will focus on one important series of clinical trials, the use of chemotherapy in treating breast cancer, and I will argue that the problems of reaching closure can be illuminated by considering them as a failure of imposing metrological standards on a diverse community.

In the first part of this paper, I discuss how cancer clinical trials in the post-war period can be viewed as an example of metrological practice. I focus on clinical trials run by Bernard Fisher, one of the most important cancer researchers of the post-war period. In the second section, I follow the controversies surrounding the claims of clinical trial researchers that chemotherapy should be given as an addition (or adjunct) to surgery in virtually all women with breast cancer, and the attempts of researchers and governmental administrators to resolve these controversies through the mechanism of consensus conferences. In the last section, I argue that the dissemination of consensus standards, like any new metrological measure, was met with resistance not least because the generalized and universalized character of those standards was incommensurate with local knowledge and practice.

The post-war multi-centre clinical trial

In a 1977 paper, 'L-phenylalanine mustard (L-PAM) in the management of primary breast cancer,' the cancer clinical trials researcher Bernard Fisher began with the hypothesis:

There is growing awareness that most if not all patients have disseminated disease at the time of diagnosis and that improvement in survival is only apt to result from employment of effective systemic therapy in conjunction with modalities used for local regional disease control.[9]

This position was markedly at odds with what Fisher labelled the Halstedian hypothesis, which had been reigning for over 75 years, namely that cancer was a local disease that spread steadily outwards from its initial location and could be cured with local and regional surgery.[10] One of the proving grounds for this new and radical disseminated-disease hypothesis, the fulcrum about which the future of adjuvant chemotherapy would turn, was the L-PAM study. This investigation is a classic example of a randomized clinical trial. It was framed with the standard clinical trial query: Is L-PAM with mastectomy better than mastectomy alone?[11] Everything in the design, execution and analysis of the trial was grounded in the predominant ethos of objectivity and lack of bias. The patients were randomized only after they agreed to enter the study, which was meant to avoid any possibility of bias. In addition, the trial was double-blinded[12] so that '[n]either the physician nor patient was aware of the treatment administered so as to ensure lack of bias in subsequent observations.'[13]

In addition, physicians could enter patients into the trial only if they believed that there was no rational basis for choosing between the L-PAM and no-drug (placebo) arms. This principle of 'equipoise' served at least two purposes. First, if physicians did not favor one arm over another, they would be less liable to unconsciously bias either patient selection or clinical evaluations. Second, physicians took the presumed equivalence of the arms as justification for enlisting patients into the trial. Indeed equipoise, unlike peer review and informed consent, was the one ethical regulation that arose directly from the practice of clinical trials. The statistical nature of a clinical trial was grounded in the null hypothesis that there was no difference between treatment arms. A statistically significant difference at the end of the trial would disprove the null hypothesis, that is, it would support one of the treatment arms. Equipoise was the ethical equivalent of the null hypothesis and helped grease the wheels of the clinical trial machine.

Fisher chose survival and disease-free survival as measures of the effectiveness of L-PAM. The former measure, although the *sine qua non* of cancer trials would not be a reliable measure for at least 10 years because of the long-term possibilities of tumor recurrence. So Fisher used survival at two years without evidence of disease (disease-free survival) as a proxy for long-term survival, a decision that would later open his studies to criticism.

In some important respects, however, the L-PAM study marked a departure from earlier cancer clinical trials. Fisher was trying to measure small differences in survival rates because he was adding a drug on top of surgery. In order to complete the study in a reasonable time, he recruited patients from a wide network of hospitals, some 82 institutions belonging to various cooperative trial groups. In addition, Fisher designed the L-PAM study as the first in a series of interlocking trials where the control arm in each study was the test regimen of the previous trial so that the studies were successively calibrated to one another through their control arms. The drug regimens were not only compared trial-by-trial, but the design also provided for a long-term drug escalation study where patients would successively receive more toxic drug regimens until a maximum tolerable combination was reached, at which point, the highest possible survival rates would have presumably been attained.

To administer his program Fisher used a centralized and hierarchical structure, with a statistical centre controlling all of the satellite institutions. The statistical (or coordinating) centre was chaired by Fisher and staffed with statisticians, data managers, secretaries and business administrators.[14] It had the infrastructure to administer the L-PAM study, and to provide the continuity and the resources required for developing, funding and pursuing further clinical trial studies.[15] This centralized body produced all clinical trial protocols and disseminated them outwards to the various investigators at the participating hospitals. The coordinating centre monitored the trial procedures to assure that the patients met the entrance requirements and that their diagnoses, treatments and assessments were conducted according to the protocol. Even randomization was centralized.[16] While all instructions flowed outward, all patient data moved back to the statistical centre where it was transferred into a computer data base system.

Although there was tight control, the trials operated effectively because of the built-in flexibility of the protocols, which could be subtly refined as they passed through the hands of a heterogeneous group of physicians, technicians, and nurses. At the same time, the overall integrity of the project was maintained by the coordinating centre, which monitored a core set of study parameters, while the participating physicians provided the glue to hold the program in check. The professional interests and political forces that led physicians to enter patients and run the studies according to protocol had to be quite robust since their clinical prerogatives and responsibilities were diminished in the multi-centre culture where they were excluded from any knowledge of the progress of the trial.[17] Indeed, the statistical centre was expressly organized to provide a 'separation of patient care and evaluation functions' because of the possibility of 'bias if study physicians are permitted access to the data during the course of the trial.'[18]

The success of Fisher's program required him to recruit physicians from successively wider domains, where practitioners at the outermost domain would serve a critical role of enlisting, treating and following the progress of patients, and record the results in reports filed with the coordinating centre. Yet, recruitment of patients was a problem that affected the L-PAM trial as well as other clinical trials. For example, it took over two years for the L-PAM trial to accrue 370 patients, which clearly attests to the reluctance of the surgeons at that time to enter patients into chemotherapy trials. Indeed, the great difficulty clinical trial researchers had in recruiting patients led to a change in strategy in the US by the late-1970s. Non-academic community hospitals, which previously had not participated in clinical trials, were brought into the cooperative networks in order to recruit more patients.[19]

Fisher's 1977 publication claimed a victory for L-PAM and forecast that even better results lay ahead with more aggressive chemotherapy regimens. Fisher based his conclusion on the finding that the disease-free survival at two years was 76.2 percent with L-PAM and 68.4 percent with placebo at the $P = 0.009$ level.[20] Nevertheless, there was a disquieting note. The gains were predominantly a consequence of the effectiveness of the drug in a stratified sample of patients under 50 years of age, while it was the over 50 cohort that was of most importance.[21]

Fisher reported on the complications of the treatment quite differently from survival. Instead of a strongly quantitative assessment, he argued more from a qualitative perspective that complications were 'limited' and 'manageable' and that there was a high compliance of the patients to the drug regimen. Although '40 percent [of the patients] on L-PAM... experienced some degree of nausea and vomiting,' Fisher characterized the complications as minimal, even though in 11 percent 'the symptoms were greater.'[22] The highly asymmetrical characterization of survival and complications typified the knowledge claims produced from clinical trials. On the one hand, the measure of success, the unit for comparing one treatment to another, was survival and its surrogate, disease-free survival. Almost the entire structure of the study, the whole of the statistical apparatus was designed to ensure that the reported survival differences were significant and not a consequence of hidden bias. Sophisticated statistical methods were developed to address pre-randomization, post-study stratification, and a host of other methodological difficulties, always with the goal of rooting out bias in reported survival. On the other hand, the analysis of complications had no such elaborate statistical paraphernalia to support it. Complications were presented as a stepchild of survival and characterized with qualitative terms like minimal and acceptable. This privileging of survival in the design and execution of clinical trials, however, as we shall see below, provided a limited measure for translating trial results into local practice.

Controversy and the road to closure among cancer researchers

Even exemplary trials, such as Fisher's L-PAM study, which was meticulously designed, delivered and thoroughly analyzed, were subject to endless criticisms. The critics of clinical trials inevitably point to their numerous limitations in the design and execution, for example, the patient groups were not appropriate proxies, or the clinical measures of success such as disease-free survival were not useful surrogates for long-term survival. Other criticisms may point to the more generic problems; for example, the small survival differences between the arms make it difficult to determine a statistically significant difference without very large and long-term studies. Or the trials are ethically wrong, and so on.[23]

The difficulty of bringing these endless arguments to closure became especially worrisome in the US by the mid-1970s. The euphoria surrounding the 1971 Cancer Act, which had pumped increased funding into cancer research, had waned and congressional critics of healthcare pressed the National Institutes of Health (NIH) to take the lead in finding a way to resolve the controversies surrounding the claims of clinical trials and the assessment of new medical technologies. Donald Fredrickson, who headed the NIH at that time, worried that if the medical community could not replace its 'informal but often haphazard process for creating authority by increment,' then outsiders might set up 'creations of "technology management," which may rely unduly on regulatory measures, or marketing controls.'[24] Fredrickson feared that if medicine's knowledge claims continued to be highly contested and if the acceptance of new techniques remained so grudging, the medical community would relinquish some of its independence to governmental agencies. As a result, he created a special branch at the NIH[25] to facilitate consensus development.[26] This program became a model for later developments of technology assessment throughout much of Western Europe. In particular, the consensus program of the National Cancer Institute (NCI) that began in the late-1970s consisted of highly structured two and one-half day proceedings where an invited panel of experts produced a consensus statement after hearing presentations by scientists and other advocates as well as comments from an open forum.[27] This consensus process, in keeping with the ethos of unbiased clinical trials, was designed so that 'neutrality is strived for in the makeup of the consensus panel, and in particular, in selecting its chairman.'[28]

In spite of the introduction of consensus conferences, knowledge claims based on clinical trials remained highly contested. To appreciate these difficulties, I will follow a particular example: the controversies surrounding the use of chemotherapy as an adjunct to mastectomy (or local excision and radiotherapy) in treating node-negative breast cancer patients.[29] Indeed, contemporary standards of practice in the US were the result of previous and very contentious debates. In 1980 an NCI consensus panel reviewed the role of chemotherapy in breast cancer and did not recommend adjuvant treatment, even though the panellists were aware that approximately 30 percent of the patients would subsequently develop metastases. By 1992, the St Gallen's Conference, a meeting of medical, surgical and radiation oncologists, recommended that virtually all node-negative patients should receive some form of adjuvant treatment.[30] This conclusion was reached in spite of the fact that during the intervening years no clinical trial had demonstrated increased *survival* with adjuvant therapy. How then did we get from the 1980 NCI Consensus Conference with no adjuvant therapy to textbook decision trees with most of its branches ending in chemotherapy or hormonal therapy?

To be sure, there were a number of studies beginning in the early-1980s that demonstrated increased *disease-free* survival. Five of them are often quoted as the most important: Milan IV, NASBP B-13 and 14, Intergroup 0011 and Ludwig V.[31] But, it was not possible to cite the results of those studies without raising new debates, about the differing protocols, patient entrance requirements, numbers of patients, country of origin of the studies, and especially whether disease-free survival was a legitimate proxy for survival.

A second NCI conference in 1985 also recommended against adjuvant therapy.[32] Since Milan IV and Ludwig V had ended by 1985, early results on the efficacy of adjuvant therapy were available as was a meta-analysis (see below) of extant trials, but this was still not sufficient to swing the tide towards chemotherapy.[33]

By July of 1988, the debate was ratcheted up when the NCI issued a clinical alert of three American trials, which reported increased *disease-free* survival with adjuvant therapy from 5 percent to 15 percent at approximately four years. The head of NCI, Vincent DeVita, argued that these findings were too significant for patient treatments to be held hostage to peer review since 'the hormonal and chemotherapy treatments described represent credible options worthy of careful attention.'[34] To obtain the approval of the investigators for early release of their results, DeVita had convinced the editor of the *New England Journal of Medicine* to drop its no publication rule for studies previously released to the news media.[35] The *New England Journal* later published the studies with commentaries by W.L. McGuire and DeVita and a number of letters to the editor.

McGuire and DeVita debated two issues: the appropriate way to measure therapeutic efficacy and who were the true subjects of the trials. McGuire argued for a broad measure of efficacy by raising questions about toxicity, and the long-term effects of treatment and costs, thereby aligning himself with the 70 percent of the patients who would not recur, but who would nevertheless receive toxic therapy.[36] DeVita, on the other hand, focused on the 30 percent of the patients who would recur without adjuvant chemotherapy.[37] According to DeVita, toxicity and cost were minor issues compared to the problem of tumor recurrence – it was survival and not complications that mattered. DeVita argued that while '[t]oxicity and the effect of these treatments on fertility ... are of concern ... so is the morbid effect of a recurrence ... and the 'toxicity' of a premature death.'[38] Throughout the clinical alert and subsequent arguments, DeVita's aims were to narrow the debate and force closure. But he unleashed a storm of controversy, which only raised new debates. Nevertheless, the early release forced the opponents into a defensive posture and put considerable pressure on physicians who became concerned about the medico-legal implications of ignoring the NCI alert.[39] Many physicians were also upset that NCI was trying to control their decisions – which in essence it was; the alert says as much.

The NCI held yet another consensus conference in 1990, in part to try to settle the controversy surrounding the clinical alert. The conference addressed a number of issues including the role of mastectomy versus lumpectomy combined with radiation in early stage breast cancer and the role of adjuvant therapy in node-negative patients.[40] The conclusions of the conference were important for two reasons. First, lumpectomy with radiation was deemed superior to mastectomy, which was a major victory for radiation oncologists. Second, and important for our story, is that the recommendations of the DeVita's clinical alert were reinforced, namely, that adjuvant therapy should be seriously considered in node-negative patients. The conference was, however, criticized for having not been definitive enough. The *Washington Post* of 26 June 1990 stated that the 'panel of experts offers little guidance to women.'[41] But that position underrated how much the conference had swung the debate towards the adjuvant therapy camp. The recommendation that chemotherapy should not be given in 'patients with tumours 1 cm or less' was by default an endorsement for adjuvant therapy in all other patients who were at enough risk that 'the decision to use adjuvant treatment should come after a thorough discussion with the patient.'[42] In fact, the low risk (one centimetre or less) group had never been a point of contention even for the strongest advocates of adjuvant treatment. In 1992, the St Gallen's conference finally issued the positive version of the NCI consensus, namely that adjuvant therapy should be given to everyone except those in the 'low' risk category.[43]

Since the early-1990s consensus statements were based on trial results similar to that available by the middle of the previous decade, why was closure reached then and not earlier? To begin with, a data synthesis, a so-called meta-analysis by an international ad-hoc group of medical oncologists was able to demonstrate that in the broadest sense adjuvant therapy in the treatment of breast cancer led to improved survival.[44] The ethos underlying the meta-analysis was similar to that of clinical trials, namely it was a comparison of treatments using measurable clinical endpoints, statistical analysis and scrupulous removal of all known bias. The analysis was carried out under well-defined protocols, which included entrance requirements that individual clinical trials had to meet, especially that they provided unbiased estimates of the study questions. Once these criteria were met, the trial was closed to further scrutiny and became a single data point in the meta-analysis, identified by an acronym like Milan-IV. All the details of the individual trials, all the messiness, the imperfections and the battles over worthiness were put to rest once the trial became a valid member of the meta-world of clinical trials. That world was, however, markedly different from the one that existed prior to the meta-analysis. To begin with, all non-randomized trials were excluded, which meant that a significant majority of the trials that had been tapped for prior debates were no longer in the running. Second, unpublished randomized trials were meticulously sought out and included in the study to eliminate the effect of publishing bias.

Meta-analysis was like an idealized consensus conference, where all the contestants were at the table, each question was represented by a quantitative result like survival, and consensus was achieved when the difference in survival for competing therapies rose to the level of statistical significance. Nevertheless, the meta-analysis by necessity greatly simplified the clinical questions. The only measures of outcome were those 'endpoints that are objectively measurable, such as overall survival,' while toxicity, quality of life and similar outcomes were explicitly excluded since they differed 'from study to study depending upon the mechanisms for follow-up and patient assessment.'[45] A further simplification was the homogenization of the patient groups, which was necessary to provide enough data to yield highly significant results.[46] The answers from the meta-analysis were definitive about survival, but vague about which chemotherapy and patient subgroups would benefit.

Because of the simplifications, the meta-analyses for node-negative patients could not be used to affect closure anywhere except in an idealized meta-world. Nevertheless, the statistical power of the results and the scrupulous attention to removing bias carried sufficient authority so that opponents could no longer argue that adjuvant chemotherapy in node-negative patients had never been conclusively shown to be effective in any breast cancer patients. Although meta-analysis could not identify which subgroups of node-negative patients would benefit, the debate had been shifted to when and in whom, rather than whether to give chemotherapy.

Indeed, it is not clear why closure was reached in the early-1990s, and we can only speculate about some contingent events that may have contributed to it. If nothing else, the pressure on oncology researchers in the US to provide definitive guidance to women was very strong by the early-1990s. The arguments for and against chemotherapy in breast cancer had been going on since the 1970s and continued to undermine the image of the medical community. And there is little doubt that continuing research support depended upon the medical community demonstrating success in certain critical areas, and especially in breast cancer therapy. Moreover, DeVita had raised the stakes with the clinical alert. On the one hand, the clinical alert was meant to demonstrate to congressional critics that chemotherapy had within its grasp one more 'cancer victory,' and on the other hand, it placed additional pressure on the cancer researchers to reach a definitive consensus. Perhaps, too, the design of the 1990 consensus conference may have encouraged a *quid pro quo* between the chemotherapy and radiation oncology communities. The recommendation that radiation with local excision should be used in preference to mastectomy in early breast cancer was a major victory for the radiation oncologists. Although the gains for chemotherapy were not as clear-cut, the recommendations of the clinical alert were reiterated and the role of prognostic indicators, which had been used as an argument against chemotherapy, was defused by recommending further research. The political stakes at a consensus conference were best put by DeVita:

> The disproof of a therapeutic hypothesis may mean the shift in management of an entire disease from one medical specialty to another. This change is generally not well received in medicine. Few clinicians appear willing to design their specialty out of a clinical experiment.[47]

The knowledge claims on behalf of adjuvant chemotherapy, as we have seen, were not the necessary result of well-designed unbiased clinical trials. Epistemic, social and political issues all contributed to the consensus that was eventually reached.

If we look at the post-war landscape of clinical trials, we find it littered with larger and ever more carefully designed studies. As the trials grew in size and number and as they influenced ever-larger cohorts of individuals, the battles over the epistemic claims of clinical trials only increased. To counter the never-ending battles, new institutions arose such as consensus conferences, clinical trial and meta-analysis cooperative groups. But the growing number of actors that were brought into the battles only increased the grounds on which medical knowledge became contested. The problem was further compounded as attempts were made to change local practices.

Changing local practice

In the physical sciences most problems are tackled by no more than a handful of research groups and an even smaller number may be involved in confirming (or contesting) the findings, while if clinical trials are to have an impact on patient treatments, they need to change the practice of every physician in the specialty. This creates serious problems for researchers in disseminating their trial results to change local clinical practice.

Again, metrology provides a useful lens to view these efforts for a number of reasons. To begin with, we are forcefully reminded that at the most generic level, communities appear almost inevitably to resist distant and powerful authorities whenever they try to impose a new order and change local customs. Second, metrology, with its focus on standards and their control leads us to appreciate that new standards often demand changes in practice, and thus local concerns must be overcome. For example, if clinicians are pressed to adopt new treatment methods, they and their patients may resist change since there may be increased risks. Third, new standards may not be commensurate with current procedures and thus may not readily translate into terms that are consonant with local medical practices. For example, the claims of cancer clinical trials with their almost exclusive emphasis on survival may not fit community needs, which may take more account of the complications of therapy and the general well-being of patients. Fourth, through the perspective of metrology we can appreciate some continuity between the ethos of clinical trials and the models researchers and government administrators and researchers used to try to align local practices.

While standards may be disseminated in the physical sciences through artefact or intrinsic standards, medical consensus standards are only propagated via protocol. There exist no material objects that can be passed from one clinic to another to assess whether a given treatment technique has been properly calibrated against the consensus conference standard. The recommendations of consensus conferences in the 1980s and onwards were codified in rules that represented an intrinsic standard, which, if applied to a population of patients, would realize a unit of improvement in relative survival. This process, like all metrological procedures, was disseminated in one direction only. Clinicians throughout the community were expected to modify their medical practice according to those standards. The successful dissemination of such standards depended, as we shall see below, on the balance of generality and specificity of the protocols.[48] We already commented that for multi-centre clinical trials the protocols had to have the flexibility in order to be adopted in various clinics. However, the overall control of the study was still maintained through the co-ordinating centre, which audited certain aspects of the protocol. In the more general problem of disseminating consensus standards through a large medical community, there is no coordinating centre so that the adoption and maintenance of medical procedures is left in the hands of local clinicians, and this again speaks to the need for carefully constructed consensus standards.

The hubristic stance of researchers that local practice could be aligned simply by disseminating standards was to a significant degree a product of the large-scale studies that began during World War II. The proto-chemotherapists had surely conquered a number of infectious diseases, and when they turned their gaze to cancer using large-scale drug screening programs they were able to demonstrate impressive tumor responses, at least early on. Furthermore, if cancer researchers like Fisher could align the practices of a wide network of physicians participating in clinical trials, then medical researchers certainly believed that consensus conferences, once they defined best practice, would only need to disseminate those standards to change medical practice. In fact, the medical community, and especially the NIH, was distressed to learn beginning in the late-1970s that nothing of the sort was occurring. Medical practices were not changing following consensus conferences. Initially, officials at the NIH Consensus Development Program believed that the problem was merely due to the poor dissemination of the standards. They attempted to insure that physicians received consensus statements through improved mailings and more carefully drawn statements that were consistently published in the *Journal of the American Medical Association*.[49] But it soon became apparent that the problems were more intractable. In one of a number of studies funded by NIH, Kosecoff and colleagues concluded that even though the findings of the four consensus conferences they investigated were well disseminated, they 'had no effect on physicians' practice.'[50] In a later paper, the authors reviewed eight studies, and used hospital records to assess the influence of consensus conferences on changing local practice. They found that in only one of the studies (caesarean childbirth) did a consensus conference have any effect whatever on modifying medical practice.[51]

If we return to the dissemination and acceptance of adjuvant therapy in primary breast cancer, we find a similar difficulty. Broadly construed, breast cancer treatments changed from the early-1970s when a diagnosis of breast cancer inevitably meant radical mastectomy to the mid-1990s when increasing numbers of patients received lumpectomy with radiation and adjuvant chemotherapy. Nevertheless, the changes in cancer therapy appeared to not have been directly correlated with the NCI and other consensus statements. The changes that occurred were slow, grudging and punctuated by large local variations. When Donald

Fredrickson developed the consensus conference concept at the NIH in the late-1970s, he felt that NIH would solve the problem of the slow increments of change that he deplored, and that clinical trials and consensus conferences would lead to more rapid and uniform changes in medical practice. What occurred was uneven change due to the resistance of physicians to adjusting their practices and bringing them in line with disseminated standards. And patients themselves also became an important factor. For example, educated women in Northeast US cities in the mid-1980s opted for lumpectomy with radiation, in preference to mastectomy alone, well before ten-year survival rates were available. These women were inclined to choose the less mutilating lumpectomy, even if there might be an increased risk of recurrence. In some mid-western communities, however, the balance remained with mastectomy even after the NCI's guidelines supporting lumpectomy with radiation were issued.[52] Indeed medical practice was and still remains a local phenomenon overlaid by general principles, which circumscribe a broad range of options for any individual patient.[53]

The resistance of physicians and patients to the disseminated protocols of standardizing bodies were a consequence of a number of problems, which arose from attempts to impose new measures on local communities. First, and at the most generic level, physicians resisted the claims of distant and powerful authorities. They saw them as attempting to impose a new order and change local custom[54] and that the researchers had 'biased and vested interests.'[55] Second, physicians would often have to learn new and more intricate surgical techniques or use regimens with more toxic drugs that, at least initially, placed their patients at increased risk and exposed them to legal redress.

Third, the protocols for new therapeutic measures were designed for generalized categories of patients, and physicians found them difficult to interpret and to apply to specific cases. Medical facts, as we saw with the adjuvant trials, were constructed by 'scaling up' the consensus statements, that is, by using more general categories of patients in order to reach statistical significance. But clinicians had to scale them back down to apply them to individual patients, which often led to difficulties. The problems physicians faced in adopting consensus statements have been amply documented in a number of studies. David Kanouse and Itzak Jacoby, who investigated the difficulties of changing local practice, concluded: 'unless a recommendation prescribes defined actions to be taken in defined circumstances for specific subgroups of patients it is unlikely to have a uniform or measurable effect on practice.'[56]

There is a fourth and perhaps deeper reason for the difficulty clinicians had in accepting consensus standards. The measure of outcome that guided physicians in clinical practice and which provides a measure for local consensus was often at odds with the endpoints used by researchers in clinical trials and consensus conferences. Cancer clinical trials often boiled down the rich and complex brew of cancer medicine into one extract, survival, while clinical practice, an alloy of the risks and benefits, concerned itself with the possible outcome in a particular patient. Outcome, in the sense that I mean here, unlike survival, is not a measurable and universal endpoint. Rather, it is some mixture of remission and cure combined with consideration of the toxicity and quality of life of the treatment.

In this respect, researchers and physicians speak different languages. The former appeals to objective measures with almost universal applicability while the latter works within a local culture with its own expectations and its own balance in reaching clinical judgements. In this sense, local medical practice is closer to metrology in the pre-modern world where it was quite common for measures of value to combine apparently disparate quantities. For instance in the middle ages, so that agricultural land could be compared and traded, the 'size' of a plot combined physical dimensions and productive capacity through the measure of the amount of seed sown in a day.[57] In like manner, clinical outcome brings together apparently disparate measures like survival, complications and quality of life. Clinicians using such measures of treatment efficacy can operate their practices within the local medical community, as long as their results lie within accepted boundaries. This commonalty of trade between physicians ties the local community together so much so that consensus statements alone cannot readily unglue it.

Although consensus conferences did not lead to the rapid and universal change envisioned by medical researchers, we should not conclude that attempts to reach such a state have not continued. During the 1990s, a new discipline of Health Technology Assessment (HTA), which has developed primarily in Western Europe and the UK, has sought to provide government officials in healthcare management, funding agencies, administrators, physicians and the public with synthesized accounts of the economic and human costs and benefits of new technologies and proposed therapies. The HTA ethos draws on the work of consensus conferences and meta-analyses. Their allegiance to objective endpoints and eliminating all known forms of bias has provided them with the respectability and trust that they need to influence such diverse groups. HTA practitioners have sought to change medical care so that it would adhere to more rationale and universal practices.[58] To influence diverse groups, these new metrologists have tried to apply marketing strategies where knowledge is packaged in various forms and disseminated to the different communities. 'A key feature of dissemination activities has been to tailor the message and choice of media to the target group's needs and interests.'[59] These target groups consisting of clinicians, administrators and policymakers should each receive different messages that would be constructed to elicit the desired response. While allegiance to objectivity and the absence of bias provides these new metrologists with the mantle of truth, the production of medical knowledge is explicitly acknowledged as an expression of social, economic and political agendas.

Notes

1 B. Latour, Bruno, *Science in Action: How to follow scientists and engineers through society* (Cambridge, Mass.: Harvard University Press, 1987), p. 251.

2 E.P. Thompson, *Customs in Common* (New York: New Press, 1991), p. 217.

3 Latour, *Science in Action*; S. Schaffer, 'Late Victorian Metrology and its Instrumentation: A manufactory of ohms,' in R. Bud and S.E. Cozzens (eds), *Invisible Connections: Instruments, institutions and science* (Bellingham, WA: SPIE Optical Engineering Press, 1992), 23–56; J. O'Connell, 'Metrology: The creation of universality by the circulation of particulars,' *Social Studies of Science*, 23 (1993), 29–73.

4 W. Kula, *Measures and Men* (Princeton: Princeton University Press, 1986).

5 Kula, *Measures and Men*, p. 48.

6 O'Connell, 'Metrology,' p. 153.

7 See e.g. C. Lawrence, 'Incommunicable Knowledge: Science technology and the clinical art in Britain, 1850–1914,' *Journal of Contemporary History*, 20 (1985), 503–20.

8 See e.g. S. Timmermans and M. Berg, 'Standardization in Action: Achieving local universality through medical protocols,' *Social Studies of Science*, 27 (1997), 273–305.

9 B. Fisher *et al*, 'L-Phenylalanine Mustard (L-PAM) in the Management of Primary Breast Cancer: An update of earlier findings and a comparison with those utilising L-PAM plus 5-Fluorouracil (5-FU),' *Cancer*, 39 (1977), 2883–903, p. 2884.

10 B. Fisher, 'Laboratory and Clinical Research in Breast Cancer – A Personal Adventure: The David A. Karnofsky Memorial Lecture,' *Cancer Research*, 40 (1980), 3863–74, p. 3867.

11 A.B. Hill, 'The Clinical Trial,' *Brit. Med. Bull.*, 7 (1951), 278–82, p. 280.

12 In this respect the L-PAM study was unusual since in most cancer clinical trials it is not possible to do double-blinded studies.

13 Fisher, 'L-PAM,' p. 2885.

14 C.L. Meinert, *Clinical Trials: Design, conduct and analysis* (Oxford: Oxford University Press, 1986), chapter 5.

15 J.D. Cox, 'Brief History of Comparative Clinical Trials in Radiation Oncology: Perspectives from the Silver Anniversary of the Radiation Therapy Oncology Group,' *Radiology*, 192 (1994), 25–32.

16 B. Fisher, 'Clinical Trials for Cancer,' *Cancer*, 54 (1984), 2609–17, p. 2613.

17 S. Hellman and D.S. Hellman, 'Of Mice But Not Men: Problems of the randomized clinical trial,' *N. Engl. J. Med.*, 324 (1991), 1565–89, p. 1586.

18 Meinert, *Clinical Trials*, chapter 5.

19 C.B. Begg *et al*, 'Participation of Community Hospital Clinical Trials: An analysis of five years experience in the Eastern Cooperative Oncology Group,' *N. Engl. J. Med.*, 306 (1982), 1076–80, p. 1076.

20 Fisher, 'L-PAM,' p. 2889.

21 Fisher, 'L-PAM,' p. 2897, table 7.

22 Fisher, 'L-PAM,' p. 2899.

23 W.W. Lawrence, 'Some Problems with Clinical Trials,' *Arch. Surg.*, 126 (1991), 370–8, pp. 373–8.

24 D.S. Fredrickson, 'Seeking Technical Consensus on Medical Inventions,' *Clin. Res.*, 26 (1978), 116–17, p. 116.

25 The Office of Medical Applications for Research (OMAR).

26 S. Perry, and J.T. Kaberer, 'The NIH Consensus-Development Program and the Assessment of Health-Care Technologies,' *N. Engl. J. Med.*, 303 (1980), 169–72, p. 169.

27 I. Jacoby, 'The Consensus Development Program of the NIH: Current practice and historical perspectives,' *Int. J. of Technology Assessment in Health Care*, 1 (1985), 420–32, p. 421.

28 Jacoby, Consensus Development, p. 427.

29 M.D. Abeloff *et al*, 'Breast,' in M.D. Abeloff *et al* (eds), *Clinical Oncology* (New York and London: Churchill Livingstone, 1995), 1617–714, p. 1673.

30 J.H. Glick, *et al*, 'Adjuvant Therapy of Primary Breast Cancer: Closing Summary,' in H.J. Senn *et al* (eds), *Adjuvant Therapy of Breast Cancer IV – Recent Results in Cancer Research*, Vol 27 (Berlin: Springer-Verlag, 1993), 296–7.

31 Abeloff, 'Breast,' p. 1672.

32 NIH Consensus Conference, 'Adjuvant Chemotherapy for Breast Cancer,' *JAMA*, 254 (1985), 3461–3, p. 3463.

33 UK BCTSC/UICC/WHO, 'Review of Mortality Results in Randomized Clinical Trials in Early Breast Cancer,' *Lancet*, 2 (1984), 1205.

34 G.A. Omura, 'Clinical Cancer Alerts: Less than wise,' in C.J. Williams (ed.), *Introducing New Treatments for Cancer: Practical, ethical and legal problems* (Chichester: John Wiley, 1992), 421–36, p. 423.

35 Omura, 'Review of Mortality,' p. 424.

36 W.L. McGuire, 'Adjuvant Therapy of Node-Negative Breast Cancer,' *N. Engl. J. Med.*, 320 (1989), 425–7, p. 426.

37 V.T. DeVita, 'Breast Cancer Therapy: Exercising all options,' *N. Engl. J. Med.*, 320 (1989), 427–9, p. 428.

38 V.T. DeVita, 'Letter to the Editor,' *N. Engl. J. Med.*, 320 (1989), 472.

39 L.R. Prosnitz, 'Medical and Legal Implications of Clinical Alert,' *J. Nat. Cancer Inst.*, 80 (1988), 1574.

40 NIH Consensus Conference, 'Treatment of Early Stage Breast Cancer,' *JAMA*, 265 (1991), 391–4.

41   Omura, 'Review of Mortality,' p. 428

42   NIH Consensus Conference, 'Treatment,' p. 394.

43   J.H. Glick *et al*, 'Adjuvant Therapy of Primary Breast Cancer: Closing summary,' in Senn, *Adjuvant Therapy*, p. 296.

44   For an overview see R.D. Gelber and A. Goldhirsch, 'From the Overview to the Patient: How to interpret meta-analysis data,' in Senn, *Adjuvant Therapy*, pp. 167–75.

45   R.D. Gelber and A. Goldhirsch, 'The Concept of an Overview of Cancer Clinical Trials with Special Emphasis on Early Breast Cancer,' *Journal of Clinical Oncology*, 4 (1986), 1696–703, p. 1700.

46   Early Breast Cancer Trialists' Collaborative Group, 'Effects of Adjuvant Tamoxofen and of Cytoxic Therapy on Mortality in Early Breast Cancer: An overview of 61 randomized trials among 28,869 women,' *N. Engl. J. Med.*, 319 (1988), 1681–92, p. 1681.

47   V.T. DeVita, 'The Evolution of Therapeutic Research in Cancer,' *N. Engl. J. Med.*, 298 (1978), 907–10, p. 910.

48   For the need for flexibility in the clinic see Timmermans, 'Standardization in Action,' 273–305.

49   I. Jacoby, 'Update on Assessment Activities,' *Int. J. of Technology Assessment in Health Care*, 4 (1988), 95–105, p. 101

50   J. Kosecoff *et al*, 'Effects of National Institutes of Health Consensus Development Program on Physician Practice,' *JAMA*, 258 (1987), 2708–13, p. 2712.

51   D.E. Kanouse *et al*, 'Dissemination of Effectiveness and Outcomes Research,' *Health Policy*, 34 (1995), 167–92, p. 174.

52   Abeloff, 'Breast,' p. 1666.

53   J. Wennberg and A. Gittelson, 'Variations in Medical Care Among Small Areas,' *Scientific American*, 246 (1982), 120–33.

54   Witold Kula remarks that 'traditional measures and ways of measuring have always been associated with sectional interests of particular social groups, and among the weak the fear of any metrological change is deeply rooted in a long experience and transmitted from generation to generation' (Kula, *Measures and Men*, p. 70).

55   A.L. Greer, 'The State of the Art Versus the State of the Science: The diffusion of new medical technologies into practice,' *Int. J. of Technology Assessment in Health Care*, 4 (1988) 5–26, p. 9.

56   D.E. Kanouse and I. Jacoby, 'When Does Information Change Practitioners Behavior?' *Int. J. of Technology Assessment in Health Care*, 4 (1988), 27–32, p. 30.

57   Kula, *Measures and Men*, p. 31.

58   H.D. Banta *et al*, 'Introduction to the Eur-Assess Report,' *Int. J. of Technology Assessment in Health Care*, 13 (1997), 133–43, pp. 135–36.

59   A. Granados, *et al*, 'Eur-Assess Project Subgroup Report on Dissemination and Impact,' *Int. J. of Technology Assessment in Health Care*, 13 (1997), 220–86, p. 226.

13
'The Best Bones in the Graveyard': Risky Technologies and Risks in Knowledge

Sally Wyatt and Flis Henwood

Introduction

> [Doctors] all want you to be on HRT [hormone replacement therapy], it's like giving a two year old a smartie. According to [my doctor] there were no side effects but I read in [the local newspaper] a few weeks ago ... that although HRT has always been to have thought to protect your ovaries now there are signs that it can cause cancer of the ovaries. So, you see, it's quite a new kind of treatment and ... people don't know what the side effects are ... They know the pros of it, they don't know the cons, but you see [medical professionals] get tunnel vision: no osteopororis, less fractured hips, less surgery. Wow! Get all these women on it but I could have the best bones in the graveyard.

Kathy told us this in December 2001. She also told us she that she had had a blood test for Hepatitis B in 1993 that revealed some hormone imbalances, so she started taking HRT orally. She stopped taking it in 1996 prior to having surgery for a suspected ovarian cyst. She started again in the summer of 2000, when she began experiencing severe hot flushes, but this time with patches. Kathy has never used contraceptive pills and was reluctant to take HRT, thinking both interfered with the body's natural cycles. As the quote illustrates, Kathy is aware that HRT can help prevent against osteoporosis but she is worried about the longer term effects. She would not have taken HRT[1] simply out of fear of osteoporosis as she feels you cannot live your life according to what may or may not happen in the future.

Kathy usually tries to ignore any health problems, but if they persist she goes to her GP. She has a wide range of resources to draw upon: she visits alternative healthcare practitioners; she reads newspaper articles and watches television programs if she happens to notice them; she read the leaflet that came with the HRT and sent off for a booklet that was advertised there; and, she also bought a book about alternative treatments for menopause. Kathy's most important source of information and support is her colleagues at the hospital where she works.

In summer 2002, the mass media was full of stories about the partial abandonment of an HRT clinical trial in the United States, called the Women's Health Initiative study. Five years into an 8-year study, the trial of Prempro, a combined oestrogen and progesterone form of HRT, was discontinued because the incidence of breast cancer reached the upper limit of acceptability. Subsequently, researchers also noticed an increased chance of heart attacks, blood clots and strokes amongst participants. In autumn 2002, the British Medical Research Council decided to stop its own Million Women Study, after a team of international advisers said that it was, 'unlikely to provide substantial evidence to influence clinical practice in the next 10 years'.[2]

When we spoke to Kathy for a second time in November 2002, she had read about the US trials in the newspapers and seen it on TV news, but dismissed the reports because she feels Americans tend to go 'over the top' about such things. She then heard about the UK trials being abandoned and started to take it more seriously. She tried to come off the HRT patches but the hot flushes became worse so she started again. She discussed this with her pharmacist and was considering cutting the patches in half. She had not yet talked with her GP (general practitioner/family doctor), because it is difficult to get an appointment with the sympathetic woman doctor and she finds the others less helpful. She thinks GPs should provide better written information about the advantages and disadvantages of HRT. She is also worried that doctors may refuse to treat patients with osteoporosis if they have not taken HRT because when she started taking HRT in 1993 she says that her GP told her that, 'if you don't start helping yourself [by taking HRT to prevent osteoporosis], you're going to find that there's no GP that's going to bother with you'.

This account of Kathy's experiences with HRT and her attempts to engage reflexively with them illustrate the central theme of this chapter. We talked with Kathy and 31 other women, half of them more than once.[3] This enabled us to reflect with Kathy and others about how their views of the advantages and disadvantages of HRT changed over time. In particular, because of the extensive media coverage HRT received during the summer and autumn of 2002 we were also able to see how such media coverage might affect women's own reflexivity in relation to risk discourses. We argue that at any moment in time, women are balancing different sorts of risks and, over time, their understanding of what constitutes risk may change. The media has an important role to play, but, as we shall see, women's own experiences and those of their friends, neighbors, colleagues and family are often more decisive. We first review competing understandings of risk found in the biomedical and social science literature generally before turning to the ways in which women draw upon competing discourses of risk when engaging in processes of reflexivity regarding menopause and its treatments.

On risk

There has been a surge of interest in the concept of risk in recent years across the natural and social sciences.[4] Technoscientific approaches to risk, found in the natural sciences, engineering, psychology and economics, are keen to establish the objective facts of risk as defined by experts. This view is based on a deficit model of lay people's understanding: people should be given more information from experts and if this is then correctly interpreted, irrational fears will disappear and lay views or 'perceptions' will come to resemble more closely the objective understandings of experts (the 'real' risks). Social science literature draws on a much wider range of risk conceptualizations, which are generally, although not exclusively, more constructivist than the technoscientific approaches. We follow Lupton's distinction between three major approaches: the 'risk society' of Beck and Giddens; cultural and symbolic perspectives informed by the work of Douglas; and the governmentality perspective associated with Foucault.[5] The publication of Beck's *Risk Society* and the contemporaneous appearance of Giddens' *Consequences of Modernity* and *Modernity and Self-Identity* marked the beginning of a new approach to the study of risk and reflexivity within social science.[6] Reflexivity is the way in which people, individually and as members of social groups such as private and public organizations, actively monitor their actions and contexts, drawing upon the knowledge available to them.[7] For Beck, a feature of late modernity is the disappearance of what he calls 'latency', the invisibility of risks. In recent years, people have become more aware of risks as previously invisible effects become ever more visible, not only to our senses but also in the media. In the case of HRT, as more and more women take it, doctors gain more clinical experience and women themselves become aware of its effects and share their experiences with others. Beck suggests that 'risks are risks in *knowledge*, perceptions of risks and risks are not different things, but one and the same'.[8] Attention is drawn here to the knowledge that produces risk assessments, pointing to the same sense of contestation, provisionality and uncertainty found in Webster's definition of an 'innovative health technology'. According to Webster, an IHT is any health technology that significantly disrupts, challenges or redefines existing sociotechnical practices, knowledge bases and resource/investment patterns.[9] IHTs occupy a contested terrain where debates over definitions and meanings of health and illness, of social values and human identity, and of risk and opportunity are seen as integral to understanding the relationship between technological and social changes in healthcare.[10] In the case of HRT, this could include the disturbance of existing expectations and practices as well as the redefinition of social and biological boundaries and it is in this sense that we refer to HRT in this paper as a 'risky technology'. We examine women's decision-making practices around HRT and analyze if and how they experience HRT as a 'risky technology' and how they deal with 'risks in knowledge'.

Giddens distinguishes between 'hazards' or dangers, those acts of god/nature which occasionally befall people and have done so throughout human history, such as floods and meteorites; and 'manufactured risks', such as taking drugs, eating food treated with pesticides or driving a car, which are everyday features of modern industrial societies.[11] This distinction between hazards and manufactured risks is important, and later we shall see if and how women make such distinctions for themselves in relation to HRT, but let us now turn to the more cultural perspectives found in anthropology. For Douglas, as for Beck and Giddens, risk is a way for contemporary western societies to cope with danger. She is concerned to understand why some activities are perceived as risky and others are not. She is critical of the cognitive and technoscientific approaches because they overemphasize individual perceptions and ignore wider social and cultural contexts in which risk is assessed. She rejects the notion that the reason someone might engage in risky activities is because he or she does not understand the dangers but suggests it may be because the person is operating within a context in which other factors are also important, including personal preferences and social norms. As she says, '[a] refusal to take sound hygienic advice is not to be attributed to weakness of understanding. It is a preference. To account for preferences, there is only cultural theory.'[12] Despite emphasizing culture, she acknowledges the reality of risks, of poor hygiene, for example. For Douglas, Beck and Giddens, there are varying levels of tension between their acknowledgement of objective risk and their commitment to understanding the social and cultural settings in which risks need to be contextualized.[13]

The governmentality approach to risk, inspired by the work of Foucault, is also relevant. Foucault's work draws our attention to the recursive character of risk assessments.[14] Information is gathered from individuals and then aggregated in order to develop population norms which are then used to advise, regulate and discipline individual behavior. For Lupton and others, risk assessments are used to regulate the body; and individuals, drawing on this type of risk discourse and the norms produced, will adapt their own practices in order to minimize risk.[15] As expert knowledge about risk has increased so too have the strategies people deploy in order to try to avoid risk. Lupton gives the example of pregnancy but this approach could equally be applied to menopausal women, some of whom, as shall be seen later, increasingly take on the monitoring and regulation of their changing bodies.

Menopause and HRT: what knowledge and whose risks?

Menopause is a contested term, and has been the subject of much discussion in both feminist and medical literature. In biomedical terms, menopause refers to the last day of a woman's final period, and thus can only be dated after the event. Menopause is assumed to be completed when a period-free year has passed. The British Medical Association (BMA) define menopause in terms of a deficiency and HRT as a simple 'replacement' for that deficiency:

> [Menopause is] the time at which a woman stops menstruating and it is a normal consequence of the ageing process. It occurs between the ages of 45 and 55, when a woman's ovaries stop responding to follicle-stimulating hormone (FSH) and produce less of the female sex hormones oestrogen and progesterone. This drop in hormone levels brings an end to ovulation and menstruation ...
>
> Hormone replacement therapy (HRT) may help to relieve many of the symptoms that occur at the menopause by boosting the level of oestrogen in the body and reducing the production of FSH. HRT is also very effective in preventing both osteoporosis and maintaining reduced risk of cardiovascular disease.[16]

Some feminists have resisted efforts by the medical profession to medicalize menopause, arguing that it is a normal part of ageing. For example, Greer[17] vehemently rejects efforts to medicalize both the symptoms and treatments for menopause. She argues that by identifying menopause as a syndrome, 'the medical establishment was empowered to treat the 'critical phase', as a complaint in which their intervention was to be sought, rather than as an important process in female development with which women themselves would have to deal'.[18] Greer[19] is critical of attempts to replace oestrogen as a misguided attempt to 'turn the clock back', instead, she argues, the medical profession should be helping women to manage the transition to a different hormonal state. Guillemin also criticizes the use of 'replacement' when she says, 'calling this drug a hormone *replacement* enables it to be seen as something other than a drug with possible side effects, something more "natural"'.[20] The BMA represents the biomedical view whereas Greer and Guillemin draw both attention to the socio-cultural context in which women grow older. Both medical and many feminist perspectives, however, deploy notions of naturalness, normalcy and inevitability in discussing menopause itself. Roberts reviews these debates, and argues, from a constructivist perspective, for the importance of understanding HRT and women's bodies in a way that does not rely upon a reification of the natural.[21] Roberts also draws attention to the ways in which women are asked to think about their pasts (and those of close female relatives, in relation to the incidence of breast cancer and osteoporosis, for example) and their futures when deciding whether or not to take HRT. Ballard adopts a technoscientific view of risk but makes a similar point to Roberts about the temporal nature of women's decision-making.[22] Ballard distinguishes between individual and collective risk, arguing that women work with both. At a collective level, women are aware of the relationship between menopause and osteoporosis, but are more likely to draw upon their individual experiences and family histories in making decisions about whether or not to take HRT.

Definitions of menopause vary and experiences certainly do. The BMA refers to the following symptoms: hot flushes, night sweats, feelings of anxiety, dry skin, vaginal dryness and urinary infections. Our respondents mention all of these, plus irregular periods, insomnia, irritability, depression, low energy, achiness, painful intercourse, palpitations and headaches. Women often experience some of these symptoms at other times in their lives; however, the combination of symptoms, as well as their severity, lead many women to associate them with menopause. In addition to HRT, there are many alternative/complementary therapies available for alleviating menopausal symptoms and preventing the onset of osteoporosis, including herbal and homeopathic remedies as well as dietary and other lifestyle adjustments. We now turn to women's knowledge about menopause and its treatments and the risks associated with both.

Social theory in social action? Giddens' double hermeneutic

So far we have focused on the ways in which social scientists understand risk and we have seen how difficult it can be to maintain the distinction between realist and constructivist notions of risk. We now turn to ways in which social actors understand menopause, its treatments and the uncertainties surrounding them. Giddens' interpretation of the double hermeneutic in social science draws attention to two processes.[23] First, social scientists need to find ways of understanding the world of social actors. Second, social scientists need to understand the ways in which their theories of the social world are interpreted by those social actors. In this chapter, we aim to do precisely that by focusing on how women draw upon the risk discourses in circulation about menopause and HRT in order to construct risk narratives for themselves. Giddens' double hermeneutic allows us to keep in view the multiplicity of risk conceptualizations and their circulation between social science and the social world.[24] We first examine the discourses around menopause and its symptoms as this provides an important context for understanding reflexivity and risk in relation to HRT. We then turn to treatments for menopausal symptoms, especially HRT and the extent to which it is considered a risky technology and how risks in knowledge about HRT are understood.

Several of the women we talked to explicitly referred to the naturalness of menopause, as part of ageing. Kathy, introduced earlier, mentioned her belief that menopause was a natural process. Ruth uses HRT, even though she had hoped she would be, 'sailing through menopause like [her] mother did without any problems whatsoever.' She 'felt the body should go through it naturally.' When asked about what causes menopausal symptoms, Jane replied: 'It's just really the shut-down of hormones in your body. The ovaries stop producing all the oestrogen ... It's the effect of the lack of hormone that causes most of it.'

As discussed earlier, the socio-cultural and biomedical views of menopause suggest different approaches to treatment. As the earlier quote from the BMA illustrates, within the biomedical/deficiency view, what is missing – oestrogen – should be replaced. This is also how Betty sees it when asked what she knows about HRT:

> Well it's hormone replacement therapy isn't it? As you get older, obviously all the oestrogen gets thinner and your bones get weaker and things like that so it just replaces what you lose as you get older basically.

HRT is used as a general term for a range of treatments, available since the 1960s. It is often offered to women during menopause or following a hysterectomy. HRT supplements the declining levels of oestrogen produced by women's bodies as they get older. It is available in various dosages and combinations of oestrogen and progesterone. HRT can be taken orally or vaginally, worn as a patch or rubbed into the skin. Implants or inter-uterine devices can be inserted to provide release of the hormones over time. HRT can be prescribed to women experiencing any or all of the range of symptoms associated with mid-life. It is claimed that HRT will relieve all of these symptoms as well as help to prevent osteoporosis and cardiovascular disease.[25] Menopausal symptoms usually return if and when women stop taking HRT. Thus, many women may take HRT for a long time. Dangers associated with HRT include increased chances of breast cancer and blood clots. The effects in relation to dementia, stroke and other cancers remain unclear. The benefits of HRT are potentially enormous but so too are the associated dangers and uncertainties, all of which remain highly contested.

As noted above, women have access to a variety of sometimes contradictory discourses for talking about menopause. The range of information and advice women might find is enormous, and it is available from many different sources and media.[26] While many of our respondents understood that HRT is still the object of ongoing medical research, not all had the same reaction. As we shall see, some women reassessed their decisions in light of the new evidence while others interpreted the changing evidence as indicative of 'risks in knowledge'. We argue that these different strategies can be understood as reflecting women's relative engagement with, and take up of, both realist and constructivist risk discourses.

Gill has been taking HRT for over 12 years. It certainly worked for her, as she described when we interviewed her in November 2001:

> [T]he hot sweats went away, everything went away and I felt fine. I was more active, it gave me more energy and I felt fine and this is

the reason that I take it ... I understand that there's certain hormones in the HRT also prevents your bones from disintegrating which I feel is really a great help, because my mother, from being quite a sturdy and upright lady, she didn't have the benefit of taking HRT, and she just crumbled down like a little old lady. And with her age and having to sit about, the pain with having this crumbling spine – that was how she described it ... I felt that I didn't want to go through that.

Gill's own feelings of good health and energy together with her memories of her mother's pain outweigh her awareness about the increased chance of breast cancer. We spoke to Gill again in November 2002. She had read newspaper reports about the US clinical trials which alarmed her, as she had not previously realized she needed to continue taking HRT in order to prevent osteoporosis. She decided herself to stop taking HRT but felt so awful with the return of hot sweats, insomnia and lethargy that she started taking it again after one month. Despite becoming more aware of some of the risks and uncertainties around HRT, Gill draws upon her own bodily experiences in deciding to continue with it.

There are uncertainties in the biomedical understanding of HRT that may or may not be resolved through continued experimentation. The generation of women now taking HRT is the same generation for whom the contraceptive pill was first easily available. There are, at the very least, uncertainties associated with the following: different hormonal combinations; different modes of taking HRT; and alternative forms of treatment; the physical and emotional effects of ageing, including osteoporosis and breast cancer; being too informed about all of these and thus anxious about the future; and of disclosing intimate bodily experiences to healthcare professionals, family and friends. We now address how women understand and deal with these uncertainties.

The uncertainties or risks in knowledge surrounding HRT have always been present, but they have become more visible since the abandonment of the large-scale clinical trials in the US and the UK in 2002. Some of our respondents have a sense of this uncertainty while some others do not. Here we illustrate different ways of dealing with uncertainty, based on disengagement from knowledge claims, one's own embodied experience and acceptance of some future uncertainty. Twelve of the 32 women do not identify any disadvantages of taking HRT, even when prompted by the question, 'I am interested in what you know about HRT. Can you tell me what you know about what it is, how it works, its pros and cons as treatments for (the respondent's particular health condition)?' Alice has been taking HRT for six years. She claims not to know about the advantages and disadvantages of HRT, and moreover, thinks that no one else does either. This helps her to feel better about the uncertainty:

> I don't think anyone really knows, as you say, what it's actually for or what it's supposed to do or what it's not supposed to do.

We interviewed Alice again in November 2002, following the press coverage about the Women's Health Initiative and the Million Women Study. She had not heard of either nor had she ever asked her doctor about the side effects of HRT. Alice is an extreme case amongst our participants in that she does not attempt to engage reflexively in relation to decisions about her health.

We met Gill earlier; she accepts that there are advantages and disadvantages to HRT but assesses them entirely in terms of her own and her mother's embodied experiences rather than in terms of statistical probability. Gill relies primarily on her own experiences, but many other participants do engage with a governmentality approach to risk and are aware of making complex decisions in which they have to weigh up statistical probabilities in light of their own experiences. Mary started taking HRT in 1997, explaining:

> The reason I took HRT is because there is a strong history of osteo-porosis on both sides of my family ... I asked my GP because I was concerned because of the osteoporosis specifically even though I knew because I'd had a thrombosis it was not necessarily the best option. And she said that in those circumstances she would rather take HRT and risk a further thrombosis rather than [risk] osteoporo-sis [in the longer term]. And I know from my own family history that my aunt who had it particularly badly was in agony.

Mary is drawing upon knowledge of her own family history as well as knowledge from an unspecified source about the link between HRT and thrombosis in order to make a decision for herself. Thus, she is behaving entirely consistently with a governmentality view in which she reflexively uses different sorts of knowledge in order to regulate and monitor her own actions in line with the expectations of the medical profession.

Liza told us that her periods had become very heavy and close together five years previously. She attended a menopause clinic, expecting to be able to have more interaction than she has with her GP. A ten-minute consultation resulted in a three-month prescription for HRT, to be taken orally. Liza says she asked lots of questions about the effects, especially as she was concerned about her sister's death from breast cancer, but she felt her questions were not answered. She received a prescription and had it filled but never took the pills. Later she asked her GP about oestrogen creams but he would not prescribe them because no one had ever asked for them before and he was not sure of their effectiveness. She was again given a prescription for pills to be taken orally which she never used. When asked about the advantages and disadvantages of HRT, Liza replied:

> I'm not sure there are any pros. I tend to think that pros tend to be cosmetic unless you have osteoporosis ... if I thought that I had osteoporosis, [that] would probably be one of the only reasons I would take it. But largely it seems to be cosmetic. It's supposed to be good for your skin and your hair, and all those other things. Cons to me – just taking chemicals. I don't really like the idea of long-term taking [of] chemicals. I think the whole thing is a con to me. I'm not sure that there are any benefits other than cosmetic unless it's to do with bone density ... People of other generations didn't use HRT and [I] guess maybe they had, maybe bone-related illnesses, but they didn't die of the menopause, did they? And people managed their symptoms without resort to HRT, and coped, managed with them more. I don't know, they just lived, just did it, I suppose.

Ultimately, the benefits do not outweigh the costs for Liza. She attempted to engage in risk assessment with both the menopause clinic and her own GP but her attempts at reflexivity were frustrated by the failure of the clinic and her GP to engage with her. Thus the opportunity for her to make the 'right' decision in line with a governmentality approach was lost so she withdraws from the process and relies on a more naturalistic approach.

Carol started taking HRT in 1996 when she began experiencing night sweats, erratic periods and mood swings. When we spoke to her in April 2002, she spoke about balancing different sorts of risk:

> [HRT] protects against heart disease, it protects your bones, *reputed* to protect your bones and just keep you on an even keel really and that appealed to me. Mind you, I don't have any concerns about my bone density ... because I breast-fed my children and I was breast-fed

... and I really don't worry about my bones, it feels solid... . Also, because I'm a smoker, and *I'm ashamed to say I'm a smoker*. But I really feel from a heart disease point of view that it's really *important to work out these kind of equations*... . I felt that since I was a smoker, that it [HRT] would help to allay the dangers [of heart disease]... . It's probably pretty silly, and I've not looked it up at all. (emphasis added)

Carol is drawing on different approaches to understanding risk. In the first line, when she corrects herself to say 'reputed to protect your bones', she draws on the idea of risk as 'risks in knowledge'. At the same time, she engages in risk assessment. Her feelings of shame at being a smoker demonstrate that she has internalized anti-smoking messages even if she has not changed her practices. She is trying to balance different kinds of uncertainty, and feels that somehow HRT and smoking might cancel each other out and that the breast feeding will help prevent osteoporosis, though she does not specify where that knowledge comes from. When we contacted Carol again in November 2002 she had not heard about the US trials but had heard a little about the UK trials. She is aware that the benefits of HRT may not be as great as first thought, but will continue taking it even though she continues to get hot flushes, particularly during times of stress. But, as she says:

I've not had any significant health problems from it, not that I'm aware of anyway, so why bother?

Carol is reverting to her own, immediate, embodied experience to justify her decision to continue despite, or even because of, being aware of the uncertainties surrounding HRT knowledge.

Jane has a history of breast cancer but she also developed severe menopausal symptoms, so faced a difficult decision about whether or not to take HRT. Together with her GP, she deals with the manufactured risks associated with HRT through engaging in further reflexive self-monitoring and possibly subjecting herself to further manufactured risks, in the form of breast cancer screening:

If you think there is something risky, you're straight to the doctor's ... you're being looked at. Whereas a woman who's not on HRT may take months and months before they actually think, 'I've got something'. So it's not necessarily that women on HRT are catching breast cancer more, it's just that they're going to their doctor more. So [my GP] said she thinks that the information is a bit corrupt because of that, it makes it look as though the women on HRT are the ones who are having breast cancer. But she said there's a lot of women out there who may have breast cancer and not be aware of it, and they're not going to their doctors, so it makes it a bit one-sided. [That] made me rethink the whole situation, because these things are only based on statistics and figures, and is that something relevant to it, maybe it's not correct at all. And then you might get something they told you is bad, suddenly they start saying, it's good... . In the end, when you've sort of gone round this circle and see fors, againsts, fors, againsts, you've just got to sit there and think, 'well what do I do?'

Jane wants to trust authority, in this case her GP and her GP's inter-pretation of medical knowledge. Like Carol, Jane experiences a tension between the risks in knowledge about HRT and the need to do some-thing. She does attempt to engage in risk assessment though her use of 'only' in her reference to breast cancer statistics indicates her awareness of the uncertainty of this type of medical knowledge. Whereas Carol disengages from the uncertainty by relying on her own bodily experi-ences, Jane, together with her GP, decides that the statistics are un-reliable and that the risks are not really very great. For Jane and her GP, the risks regarding HRT were not so much about risks in knowledge as about 'bad science'. We also interviewed Dr Carter, the GP mentioned by Jane and when we asked Dr Carter about whether patients think about HRT in terms of risks and benefits, she replied:

I think that the major concern, in spite of the media thing on car-diovascular, is still breast cancer, but of course the statistics are very misleading... . They mention things like doubling risks and so on, but of course the risks of having breast cancer are quite small anyway and it's not a gross increase in risk. And, there's a possibility that – in fact, I haven't see the papers though I have tried to see – that although there's an increased pick-up rate of breast cancer it may just be that it's picked up sooner because women that take HRT tend to be more breast-aware and therefore they examine them-selves more or get examined more frequently. And so it's picked up earlier ... and therefore at a less deadly stage.

Jane correctly recalled what her GP had told her and accepted it as fact. Understanding large-scale studies and probabilistic conclusions is not easy, neither for patients nor professionals.

We have examined the ways in which women draw upon a range of reflexive resources in order to make decisions about whether or not to take HRT. Earlier, we demonstrated how social scientists draw upon both realist and constructivist notions of risk, sometimes within the same work. Similarly, we have demonstrated that social actors, both patients and healthcare practitioners, slip between realist and constructivist notions of risk. This confirms Giddens' view of the double hermeneutic in social science. This slippage between realism and constructivism is evidenced in accounts that move between understanding HRT as a technology that carries risks that can be 'known', and understanding that there are 'risks in knowledge' surrounding HRT.

Conclusion

Many of the women we spoke to are struggling to reconcile the manufactured risks posed by drugs, uncertainties or risks in knowledge about the effects of those drugs and the very human fear of ageing. It would be comforting to think that ageing is a 'natural' and unproblematic process yet, for many of the women we interviewed, the medical situations are not uncomplicated. The range of symptoms, the prescribed treatments and the after-effects experienced vary, and the symptoms are sometimes very severe. In part, this confirms the feminist view that menopause is not easily reducible to medicalized definition or treatment. Moreover, the variety of individual experience reinforces the extent to which there are risks in knowledge about menopause and HRT, both for individual women and for the medical profession.

Our research enables us to see how women conceptualize and experience risk in their daily lives. We do not seek to promote one risk discourse over another. We are not suggesting that socio-cultural risk discourses are always to be preferred to technoscientific ones. Rather, we are arguing that there are several risk discourses in circulation, and that women draw upon versions of these in their own more and less reflexive efforts to make sense for themselves of menopause, its symptoms and its treatments. Many respondents, not surprisingly, express ambivalence in their understanding of risk. Deeper understanding of women's ambivalences and the multiplicity of risk narratives can only be achieved through analysis of everyday life experiences. Such an understanding could result in more appropriate forms of information being developed for patients which take into account socio-cultural factors as well as medical ones, and which acknowledge the uncertainty inherent in all forms of long-term treatment.

We have illustrated three different ways women have of dealing with uncertainty. First, there are those, like Alice, who profess ignorance and attempt to disengage from any sort of knowledge claims. Second, many women, such as Gill, rely on their own embodied experiences. Third, women like Mary accept some future uncertainty. As Roberts[27] suggests, many of these women, including Gill and Mary, are engaged in a complicated temporal balancing act. They are seeking relief from immediate and present symptoms, while at the same time considering the medical history of themselves and their close female relatives and the future possibility of cancer, heart disease, osteoporosis and dementia. For some women, such as Carol, the short-term relief from symptoms outweighs any unknowable future problems.

Most of the women we interviewed have a sense of HRT as an innovative and risky technology, with the potential to disrupt their own identities, their relationships with their own bodies and with other people. Some women, like some analysts, move between realist and constructivist notions of risk when accounting for their actions. Even though women may not be aware of the different and incommensurable risk discourses operating in the scientific and social scientific literature, their attempts to inform themselves about menopause and HRT often confront them with 'risks in knowledge'. There are calls, both in the feminist literature and in our participants' accounts, for the acknowledgment of the ways in which social and cultural factors shape women's feelings about risk but this call, in and of itself, should not be equated with any formal challenge to the dominance of the technoscientific paradigm in healthcare decision making. The field of consumer health informatics is replete with attempts at finding ways of quantifying consumer preferences precisely so these can become part of a formalized risk assessment process.[28] However, even when dealing with the statistics generated within biomedical and risk assessment paradigms, women and healthcare professionals struggle to make meanings relevant for their own situations. The difficulties of statistical interpretation exemplify the struggles many women have in wanting calculation and certainty in an uncertain environment. Given the extensive media attention HRT received in 2002 and 2003, some women, though certainly not all, became aware of the provisionality of biomedical knowledge and some attempted to engage reflexively with this knowledge. This provisionality, together with many women's awareness of the ways in which menopause and HRT are differently experienced, contributes to their appreciation of risks in knowledge.

Acknowledgments

The research on which this article is based was supported by the Innovative Health Technologies Programme, funded jointly by the UK Economic and Social Research Council and Medical Research Council (Project No. L218252039). Further information about the Programme can be found at www.york.ac.uk/res/iht. We are grateful to Angie Hart, Julie Smith, Audrey Marshall and Hazel Platzer for their contributions to the project.

Notes

1 The changing and contested nature of HRT is the focus of this chapter, thus it is not appropriate to provide a simple definition of HRT at this stage as we do not wish to privilege one definition over any other. A fuller discussion of HRT appears later.

2 *Independent*, 24 October 2002, p. 10

3 This research draws upon the experiences of 32 women living in South-east England who were seeking to inform themselves about menopause, and its most widely-used treatment, HRT. All had received prescriptions for HRT. Menopause is not life threatening, and symptoms can persist for a long time, thus women have time to inform themselves both about the condition and possible treatments should they so wish. Interviews included questions about their reasons for considering HRT, their understanding of how it works and their understandings of its advantages and disadvantages. Participants were also asked about their awareness and use of alternative treatments. In addition, women were asked about whether and how they looked for health information generally as well as for HRT and other treatments for their symptoms. They were asked where they look and where they find information; by what means they find it; how they interpret and make sense of it both for themselves and in negotiation with others, including in consultation with healthcare practitioners. Women were often recalling events and decisions that had been made many years previously. A second round of interviews was conducted with 16 of the women, six to 12 months after the first interview, in order to provide a longitudinal dimension to the study. Interviews were conducted between November 2001 and January 2003. Fuller details of the methodology and results can be found in S. Wyatt, F. Henwood, A. Hart and J. Smith, 'The digital divide, health information and everyday life,' *New Media and Society*, 7:2 (2005), 199–218; A. Hart, F. Henwood and S. Wyatt, 'The role of the Internet in patient-practitioner relationships: Findings from a qualitative research study,' *Journal of Medical Internet Research*, September, 6:3 (2004), e36. Available online: http://www.jmir.org/2004/3/e36/; F. Henwood, S. Wyatt, A. Hart and J. Smith, '"Ignorance is Bliss Sometimes": Constraints on the emergence of the informed patient in the changing landscapes of health information,' *Sociology of Health and Illness*, 25:6 (2003), 589–607; and F. Henwood, S. Wyatt, A. Hart and J. Smith, 'Turned on or Turned off? Accessing Health Information on the Internet,' *Scandinavian Journal for Information Systems*, 14:2 (2002), 79–90.

4 A quick search of the social science citation index (ISI Web of Knowledge, Institute for Scientific Information, www.isinet.com/isi/. Accessed 30 June 2003) confirms this. The search term 'risk' yielded 884 hits in social science articles published in 1988, 5,414 hits in articles published in 1995, and 8,509 hits for 2002. A similar pattern of growth, though with larger absolute numbers, can be observed in the science citation index, where 'risk' resulted in 2,857 hits in 1988, 22,696 in 1995 and 42,234 in 2002. In much of this latter, natural science literature, the focus is on risk analysis, assessment and management, with a view to specifying risks and the probabilities that can be attached to them.

5 D. Lupton, *Risk* (London: Routledge, 1999).

6 Beck's *Risk Society* was first published in German in 1986 and appeared in English in 1992. Giddens' *The Consequences of Modernity* appeared in 1990 and *Modernity and Self-Identity* in 1991 (Cambridge: Polity).

7 Giddens, *Consequences of Modernity*.

8 Beck, *Risk Society*, p. 55, italics in original.

9 A. Webster, 'The Innovative Health System: Implications for social science research,' IHT Programme Launch, BMA House, London (2000), www.york.ac.uk/res/iht/events/launch/andrewpp.ppt. Accessed 10 August 2004.

10 Other changes identified by Webster include: novel functionality; new production, new markets, new regulation; novel organizational demands, and redefinition of skills.

11 A. Giddens, *Runaway World: How globalisation is reshaping our lives* (London: Profile Books, 1999).

12 M. Douglas, *Risk and Blame: Essays in cultural theory* (London: Routledge, 1992), p. 103.

13 This tension is also present here, and has parallels with the exchange between V. Singleton, 'Feminism, Sociology of Scientific Knowledge and Postmodernism: Politics, theory and me,' *Social Studies of Science*, 26:2 (1996), 445–68 and 'The Politic(ian)s of STS,' *Social Studies of Science*, 28:2 (1998), 332–7) and H. Radder, 'The Politics of STS,' *Social Studies of Science*, 28:2 (1998), 325–31 and 'Second Thoughts on the Politics of STS,' *Social Studies of Science*, 28:2 (1998), 344–8 regarding the normative implications of applying actor network theory to the cervical cancer screening programme in the UK. C. Roberts also raises the difficulty faced by feminists studying biomedicine who are often confronted with the question, 'Should I take it or not?' in her article, '"Successful Aging" with Hormone Replacement Therapy: It may be sexist, but what if it works?' *Science as Culture*, 11:1 (2002), 39–60.

14 M. Foucault, *Discipline and Punish: The birth of the prison* (London: Penguin, 1977).

15 Lupton, *Risk*.

16 *Menopause*, British Medical Association, http://www.bma.org.uk/ap.nsf/Content/HAmenopause. Accessed 15 April 2004.

17 G. Greer, *The Change: Women, ageing and menopause* (London: Penguin, 1991). This book – a response to the medicalization of menopause – was mentioned by several respondents, both patients and professionals, even though it is out of print.

18 Greer, *The Change*, p. 25.
19 Greer, *The Change*.
20 M. Guillemin, 'Blood, Bone, Women and HRT: Co-constructions in the menopause clinic,' *Australian Feminist Studies*, 15:32 (2000), 191–203, p. 197. Indeed, in the US, 'replacement' has largely been dropped, and drug companies refer simply to 'hormone therapy'.
21 Roberts, *Successful Aging*.
22 K. Ballard, 'Understanding Risk: Women's perceived risk of menopause-related disease and the value they place on preventive hormone replacement therapy,' *Family Practice*, 19:6 (2002), 591–5.
23 A. Giddens, *The Constitution of Society* (Cambridge: Polity, 1984). Within philosophy, the 'double hermeneutic' is used more generally to refer to the problem that social scientists have in dealing with the interpretations of social life produced by social actors themselves as well as the interpretations of social life produced by analysts.
24 J. Tulloch and D. Lupton, *Risk and Everyday Life* (London: Sage, 2003) attempt something similar in their study of 134 people in England and Australia and their approach to risk in their everyday lives.
25 British Medical Association, *Menopause*.
26 Henwood *et al*, *Ignorance is Bliss*.
27 Roberts, *Successful Aging*.
28 G. Esyenbach, 'Recent advances in consumer health informatics,' *British Medical Journal*, 320 (2000), 1713–16.

14
The Politics of Endpoints

Stuart Blume

Faith, hope and evidence

For half a century the Randomized Controlled Trial has been taken as a 'gold standard' in establishing the value of a new drug or other technology.[1] Its introduction epitomized the hope of rationalizing medical practice, constantly reiterated by journal editors and textbook writers over the years.[2] And yet, despite all the statistical sophistication, medical practice proved more recalcitrant than some had hoped. For one thing, as Doll recounts, there was significant opposition to the notion of randomly allocating patients to one regime or the other.[3] Despite the growth in numbers of trials, in the 1970s many new technologies were still being introduced on the basis of one or two enthusiastic reports, usually written by doctors with an interest in the technology they were advocating.[4] This was especially true when a new technology seemed to promise a major breakthrough in treatment, or opportunities for research, reputation building, or commercial success. For example, when the CT scanner appeared on the market, hospitals queued up to buy an instrument that (at the time) cost more than $300,000 even though virtually nothing had been published regarding its utility in patient management.[5] Given political concerns at the rapidly rising costs of healthcare, and given that economists were pointing to new technology as a major culprit, the response of governments in many countries is hardly surprising. The attempts at rationalization that followed were largely inspired by financial concerns.

From the 1970s governments sought more control over the introduction, spread and use of expensive new medical technologies. They could do this partly by making decisions dependent on convincing demonstrations of safety and efficacy. Later the list of things that had

249

to be taken into account was extended. Did the additional benefits over previous treatments justify the (inevitable) extra cost? Some technologies were seen as having implications for the organization of medical work, or of an ethical character. The emergence of Health Technology Assessment (HTA) was also part of this response. HTA was to have enlarged the evaluation process 'to encompass not only the clinical consequences, but also the economic, ethical, and other social implications of the diffusion and use of a specific procedure or technique on medical practice. Technology assessment thus takes a broad perspective and its aim is to provide facts as a basis for not only clinical decision-making, but also for policymaking in healthcare as a societal endeavour'.[6] It didn't work out quite like that. Analysis of the potential social implications of ethically problematic technologies, like xeno-transplantation or genomics, tends to be pursued in ad hoc advisory groups outside formal HTA.[7] Though the initial ambition of HTA may have been scaled down, efforts being devoted to the rational appraisal of new techniques continue to grow. There is a widespread assumption that general conclusions can be reached regarding the benefits (if not the costs) of medical technologies. So conclusions will typically have the universalistic form of scientific laws: 'the treatment is effective with patients of this or that age having this or that symptom'. As in science, the social and cultural particularities of knowledge production are stripped away. We don't conclude, on the basis of a trial carried out in the North of England, that the treatment has been shown to work for patients in the North of England. And that, of course, is because only certain variables are assumed to be relevant: symptoms, to be sure, age and gender perhaps... but not the wealth or poverty in which people live, or the kind of work they do, or the language they speak at home. So it would be reasonable for colleagues in Sweden, or Canada, or Russia to conclude, on the basis of that study carried out in the North of England, that the treatment would be just as effective for any of their patients showing similar symptoms.

Today a lot of effort is put into aggregating results from clinical trials carried out all over the world – so-called meta-analysis – to produce what seems to be a still more authoritative answer to the question of whether an intervention works. This of course implies, among other things, that it is not necessary for health authorities in each and every country to do their own assessment. Where resources are inadequate, time pressing, or numbers of patients too small, decisions can reasonably be based on meta-analysis of trials done in other places. This, it seems to me, roughly approximates the sense of how things should be that dominates health policy in the industrialized world. It has a two-fold rationale. For physicians there is the promise that a new drug or device can be safely and effectively used in caring for patients. For policymakers analytical techniques like HTA, cost-effectiveness analysis and the rest, should provide some protection from the vicissitudes of politics. A decision grounded in objective analysis can be presented as one that follows logically and incontrovertibly from the evidence, and so one in which interest group politics and special pleading have played no role. The possibility of endless political controversy is closed off. We can all be reassured that the decision reached was a fair and legitimate one, and one that should lead to limited resources being optimally deployed. This sounds self-evidently correct. As citizens we have an interest in limited healthcare resources being used optimally... and as (potential) patients we want to have faith that the treatments being offered are the best possible ones. We want to have faith and we want to feel that our faith is well-grounded.

But there is a complication to our sense of '*how things should be*' that needs to be introduced. In Britain, as in many other countries, 'consumers of healthcare' are now taken far more seriously than they used to be.[8] What does this imply? How do patients, or 'consumers', think about their treatments, or the treatments they'd like? Often, of course, patients' ways of thinking are influenced by those of their physicians: their vocabulary becomes medicalized and their hopes of a cure reflect what doctors hold up to them. But not always. And perhaps especially not when they have learned to live with their illness or disability over a long period. A study carried out in Britain by the Alzheimer's Society found that the symptoms that most bothered patients or their carers weren't the symptoms that new anti-Alzheimer's drugs had been intended to alleviate. In clinical trials of drugs like Donepezil, the concern – the endpoint in assessments – had had to do with cognitive functioning. But patients were at least as concerned by quite other things: by their own anger and forgetfulness for example.[9] On the basis of this study the Society argued to the British National Institute of Clinical Excellence (NICE) that qualitative data based on patient experience also had to have a place in determining the value of new treatments.

At a time of growing acknowledgement of consumer rights this made sense. Patients and consumers <u>did</u> have a right to be involved in thinking about such things. The UK has become a leader in finding ways of associating healthcare consumers with formal evaluation processes.[10] In the USA patient and consumer groups have gained a voice in quite other ways than in Britain: far more through a process of adversarial politics, as studies in the fields of AIDS and breast cancer activism make abundantly clear.[11] But however it comes about, the involvement of patient/consumers in the work of reflexion and evaluation amounts to tacit acceptance of the fact that complete rationalization was never wholly feasible. The result is a compromise, between on the one hand the authority of clinical and economic 'evidence', and on the other the political negotiation of expected benefit. We can't assume that consumers' desires, expectations, experiences – what they bring to the discussion – will be determined by the clinical evidence. Indeed, everything that medical anthropology has taught suggests quite the opposite. The way people experience illness and treatment, their hopes, and the values and concepts they use in reflecting on their experience are grounded in their particular cultures and the socio-economic circumstances under which they live. 'Letting the consumer in' means that it will become ever harder to sustain the notion of an objective and universally applicable assessment of a medical technology.

There is a fundamental tension here between the rhetoric of evidence on the one hand and the expectations of medicine infusing western culture on the other. We have been led to accept that, given time and money, the achievements of biomedicine are potentially unlimited. The assumption is that progress can be sustained indefinitely into the future: what Daniel Sarewitz calls 'the myth of infinite benefit'.[12] David Rothman has argued that this belief in medial progress, and a consequent unwillingness to be denied any part of that potential, has its roots in the individualistic values of the American middle class.[13] If so its consequences affect all of us. Daniel Callahan sees medicine as the most powerful tool that individualism has at its disposal. 'In the use of medicine to control procreation,' he writes, 'to manage mood and affect, and to rescue people from the folly of their personal choices and hazardous lifestyles, modern medicine allows individualism a range of expression that would be otherwise inconceivable. The affluent can turn to medicine for lifestyle therapy, and the poor (if they can afford it) for amelioration of the effects of their poverty'.[14] In the last two decades, with awesome advances in genetics in particular, but also in immunology, in the neurosciences, in cellular biology – advances which few non-specialists can fully grasp – there seems ever more reason for faith. We are led to believe that in all respects each and everyone of us, and our children, can be normal (at least) in all senses of the word. We don't have to be unusually large or unusually small. We have acquired the right to one or two children of (at least) normal intelligence and with no more than the usual likelihood of becoming ill. We are told, and are more than willing to believe, that as we grow old our fading eyesight and hearing, our sagging flesh and weakening bones, our diminishing sexual potency, can all be restored.

Books, articles, TV programs, websites extolling the many benefits of advancing medical technology multiply. We are told about the diseases soon to be cured or prevented – AIDS, Alzheimer's disease, cancer, malaria – the genetic defects which will be reparable even *in utero*. We cheer, of course. We donate money to the charities supporting this work. The importance of continuing to invest in medical research, of not placing insuperable regulatory barriers, or financial barriers, in the way of entrepreneurial activity is clear to all of us. And yet: on the other hand books, articles, websites warning of the dangers of medical technology multiply also. Drugs are insufficiently tested for side effects. Xenotransplantation is not only ethically problematic (for some) but also poses risks of viral transfer. Drugs are being used to control the natural exuberance of children or correct for parental inadequacies. Faced with these claims and counter claims, and with soaring costs, governments have sought still better ways of establishing beyond doubt the merits of each and every new drug, vaccine, therapy, device as it appears. Now it's Evidence Based Medicine, gathering and sifting the best available data, which should help eliminate special pleading and controversy, and should lead to decisions that will find widespread support.[15]

This is a vain expectation. What stands in the way has not changed in essence over the past half century. A different ethic of medical practice is still to be found. Economic interests have become all the greater with rising R&D costs and the globalization (or transnationalization) of industry.[16] There is something more. Medical technologies, and the institutions which develop them, are the source of a hope, a consolation that people will not abandon for the sake of a few statistics. I don't want to know that the drug is only effective in 10 percent of cases, and I certainly don't want to be told that 'although it might work, although my life could be prolonged, I can't have the drug because scientists aren't 100 percent sure that it works'. Why should my chance of life be jeopardized by scientists' methodological scruples? HTA, meta-analysis of clinical trial data, cost-effectiveness analysis, and all the other tools now available, won't make decision-making regarding new medical technologies a wholly rational business because they won't stop manufacturing industry, or hospitals, or patient organizations from lobbying for what they think necessary. Not infrequently they will work together, forming a powerful coalition. The result is that the context within which evidence has to function is becoming more and more political.

Contexts of evaluation

We know from studies of drug regulation that data on efficacy and safety can lead to quite different decisions in the different regulatory regimes to be found, for example, in Europe and the United States.[17] But in the cases I shall discuss here, precisely the 'regime' or more narrowly the 'jurisdiction', in or under which decisions are made (or agreement reached), is also at stake. Conflicts over jurisdictional status are an increasingly important aspect of the interaction between biomedical science and politics today.[18]

Consider the following example.[19] A Canadian randomized trial had shown that breast cancer screening conferred a slight but non-significant benefit on women in their fifties but not on women in their forties. Despite furious attacks from radiologists on the quality of the trial, the US National Cancer Institute withdrew its support of annual screening for women in their forties, although the American Cancer society continued to advocate screening of the younger age group. The debate rumbled on. New evidence appeared. The NCI appointed a committee to review this new evidence. To the fury of many, including many radiologists, their report advised against any extension of screening to all women in their forties. The committee and its report were attacked widely and the NIH was required to defend its position before a committee of the US Congress. The Senate subsequently voted in favor of mammography for younger women. Susan Fletcher, chair of the committee that had recommended against extension of screening attributed all of this to the powerful commercial interests involved. Kaufert disagrees. Her conversations with radiologists led her to the view that it was their clinical experience, and that alone, that had convinced radiologists. Without screening women would die who could have been helped. That fact, and not population-level statistics, was for them the only appropriate test.

Many issues can be extracted from this episode. It clearly has to do with inter-speciality jurisdictional disputes. Radiologists and epidemiologists collect, and are persuaded by, different sorts of data. The studies they design will make use of different endpoints as indicating the added value of a new approach. Whose perspective will command ultimate authority? What it also highlights is the contingent nature of the forum in which agreement has to be reached (or decisions made). In this case, political pressure led to the issue being removed from the specialist jurisdiction of the NIH and decided by the US Congress. In another instance, the controversy regarding the safety of silicone breast implants, the courts played a major role, at least in the USA.[20] What counts as evidence, and the ways in which evidence can be made convincing, are by no means the same as between these different fora. Ethicists, worrying about interventions which have not yet been the subject of clinical trials, reflect on the implications their views might have. Thinking through the implications of successful somatic gene therapy, Gardner, for example, suggests that this could, in principle, open the way to non-therapeutic human genetic enhancements.[21] Bringing 'enhancements' within the scope of medicine, it is widely acknowledged, could pose profound ethical and financial problems. Hence the attempts at making a robust distinction between treatment and enhancement.[22] But widespread use of genetic enhancement could come about whatever ethicists may decide. The notion of the 'slippery slope' can be helpful here, especially if we note the distinction that has been made between logical and empirical slippery slopes.[23] A logical slippery slope may ensue when a criterion intended to demarcate what's allowed and what isn't proves too loose. An empirical slippery slope ensues when, even if clear lines of demarcation are drawn, the relevant moral distinctions won't adequately influence the choices that people make. That is what could occur here. Using various notions of competition, between parents for the success of their children, and between nations for economic success, Gardner argues that an empirical slippery slope is very likely here. Norman Daniels has expressed similar doubts.[24] In other words, whatever moral arguments and whatever demarcation criteria ethicists may advance, they may lack the authority or influence to make them stick. To use a military metaphor, it may in practice be impossible to 'hold the line'.

To whom precisely mammographic breast cancer screening should be offered, in the USA, mattered greatly to radiologists, to hospitals, to the suppliers of radiological equipment, and to a well-organized and influential constituency of women's health activists. Faced with so powerful a coalition of interests epidemiologists could not hold the line. They were unable to sustain the authority of their evidence or the conclusions they drew from it. There's an interplay, however, between the weight of any kind of evidence or argument, and the forum in which decisions are reached. The following example should enable us to explore this interplay in more detail.

Social policies and the renegotiation of endpoints

The cochlear implant is a device intended to provide totally deaf people with a kind of hearing. The idea is simple enough. Sounds are converted into electrical impulses that by-pass a non-functioning inner ear and are used to stimulate the nervous system directly. The first attempts at developing a device along these lines go back to the 1950s, though at that time the technology available was not up to the job. Necessary advances in micro-electronics and biocompatible materials came later. In Britain research had started in the early-1970s, and by the late-1980s a number of implant centers had been established, one of the first being in Manchester, England. By this time the first commercial devices were on the market, with two devices dominating.[25] One of these, based on work done by the American William House, was manufactured by 3M. The other, produced by an Australian company, the Cochlear Corporation, was based on work done by Professor Graeme Clark in Melbourne. The FDA approved the implants for use in adults in 1984/5, and the Nucleus device for use in children in 1990. Also in 1990 – after intensive lobbying – the British government announced that £3 million over 3 years would be earmarked for cochlear implant programs.

Between the end of the 1980s and the mid-1990s a number of reviews of the evidence for the benefits of cochlear implantation appeared. Most of them reached more or less the same conclusion, specifically about the implantation of deaf children.

In 1989 a comprehensive and authoritative review of progress in the implantation of children was published.[26] In a concluding chapter Dorcas Kessler, of the UCSF Department of Otolaryngology, reviews data relating to the two devices (3M and Nucleus) then approved for experimental use in children in the US. 200 to 300 children had been implanted with the 3M/House device, and around 100 with the Nucleus. The tone of the review is optimistic but critical. So far as the House/3M device is concerned, there was no clear evidence regarding 'the acquisition or maintenance of intelligible, spontaneous speech ... the implant does not eliminate the need for special services and for consistent, ongoing training'[27] [There was much less information relating to children fitted with the Nucleus, although a few seemed to have very rapid benefit]. Lack of appropriate, standardized and longitudinal data was a major problem in reaching a definitive assessment. Dorcas Kessler concludes with the remark that 'Until such findings are available, it may be appropriate to decrease the rate at which young deaf children are being implanted, so that, when implanted, it is known that they will receive the best currently available system'.[28]

The FDA's decision in 1990, approving use of the Nucleus device in deaf children, also concluded that the device provided a consistent improvement in the hearing of deaf children whose hearing was not adequately assisted by hearing aids. The FDA was convinced that sufficient assurance of the safety and effectiveness of the Nucleus Cochlear Implant, when used in children between 2 and 17 years of age, had been provided.

In 1995, reviewing experience and results obtained in the UK, Summerfield and Marshall reach the provisional conclusion that 'the implantation of children can bring material benefits which, though variable, are generally positive and are sometimes impressive... Overall, however, we judge that the best outcomes will be found when pre-lingually deafened children are implanted when very young in the context of a strong commitment to implantation, rehabilitation, and spoken language'.[29]

In May 1995 the National Institutes of Health convened a (second) 'Consensus Development Conference', to review the state of knowledge regarding the technology. The data available were quite extensive, though there was still a striking lack of longitudinal studies of the effects of implantation on language development based in contemporary neuro- and psycho-linguistics. Many of the effects of cochlear implantation in deaf children were still uncertain. Despite this, the consensus panel was convinced that the benefits of implantation in both adults and children had been demonstrated:

> Cochlear implantation improves communication ability in most adults with deafness and frequently leads to positive psychological and social benefits as well. The greatest benefits to date have occurred in postlingually deafened adults ... Cochlear implant outcomes are more variable in children. Nonetheless, gradual, steady improvement in speech perception, speech production, and language does occur. There is substantial unexplained variability in the performance of implant users of all ages.[30]

The fact that so little was known about the effects on children's linguistic development clearly weighed less heavily than the fact that it did seem to offer them improved perception of speech. In other words speech perception was effectively and implicitly assumed to be the goal towards which progress had to be measured.

Of course, by the 1990s clinical evidence alone was no longer sufficient to persuade policymakers and health insurers. There were too many promising treatments clamoring for limited resources. This of course is where economists came in. A new intervention didn't simply have to be effective, it had to be cost-effective. In other words, what it offered in terms of health outcomes had to represent a better use of available resources than other interventions competing for the same money. At the First European Symposium on Paediatric Cochlear Implantation, in 1992, Professor Mark Haggard introduced participants to the need for new kinds of assessments.[31] A politically effective – and that meant economic – case would have to be made for pediatric cochlear implantation. In the succeeding decade a number of evaluations of pediatric implantation using Quality Adjusted Life Year (or QALY) measurements have been published.

Most of the ones that I've looked at depend upon assumptions that actually entail a tacit reformulation of the purpose of the intervention, from speech perception to social functioning. For example, one study assumes that, 18 months after implantation 65 percent of children would have moved from special to mainstream education;[32] another that 50 percent of children would eventually move.[33] Without questioning the effects of such a shift on educational attainment school transfer seems to have become the *de facto* endpoint. All of them reach the conclusion that, in terms of the improvements in quality of life that the technique offers, cochlear implantation is good value for money. This assumption of a shift from expensive special education to cheaper mainstream education is crucial to the calculus. In reaching this conclusion a vast array of limitations in data, simplifications, and so on, are excised. And we can be sure that, wherever the studies were conducted, contextual factors will be stripped away as decision-makers around the globe tell each other that 'cochlear implantation is good value for money'.

All these assessments, conducted by and within essentially clinical-audiological perspectives were reaching similar, cautiously positive, conclusions. Some others, however, were looking at the same evidence rather differently.

In a book that appeared in 1992 Harlan Lane, an American Professor of Psychology and leading advocate of Deaf culture, also reviewed the outcomes literature.[34] His review stressed the enormous gaps in current knowledge, suggested methodological inadequacies, and implied that the wrong questions have been asked. Lane's critique is partly methodological and statistical. He tried to look at the findings from the point of view of a questioning parent. 'The first and fundamental question that parents have for the doctor is: Can you make my deaf child hear? They generally do not mean: Will implant surgery give him any hearing at all? They mean, rather: Will he be able to hear well enough to learn our language, to communicate with us, with his teachers, and with other hearing people? In short, they want the deafness undone'. So far as he is concerned, the clinician, reviewing the research literature, can say no more than that such an outcome is statistically unlikely, and that very little can be said regarding a particular child's likely benefit. On this reading there is no evidence that an implanted child benefits sufficiently to pass through school as a hearing child. At best, he argues, the deaf child could function as a severely hard-of-hearing child. And so far as school achievement is concerned, severely hard-of-hearing children don't do much better than deaf children do. And that, surely, is the more appropriate test.

Promoters of cochlear implantation have available a powerful collection of rhetorical tools, strengthened by their implicit universalism and their appeal to the policymaker's cost-consciousness ('the technology is acceptable value for money' or 'there is considerable potential for cost-saving'). The critique of the Deaf community, which largely opposes cochlear implantation especially of deaf-born children, appeals to a collective history of oppression, to the value of sign language in individual lives. Their arguments are frequently grounded in, and legitimated by appeal to, personal testimony. In a medical forum evidence of this kind has little weight. But in a different, perhaps more political forum, that could be a different matter.

Despite his eloquence Harlan Lane and his allies in the American National Association of the Deaf were unable to shift evaluation of the implant out of the purely medical/audiological forum. In France, two years later, that was accomplished. In May 1994 a group of linguists, social scientists, Deaf activists and parents of deaf children presented a document to the French National Committee on Medical Ethics (CCNE).[35] In it, they argued that given the uncertainties regarding the linguistic, psychological and social implications of implanting deaf children, the technique should be regarded as experimental. Under French law, this would subject its use to rigorous control and oversight. The Committee, a high-status body independent of any specialist interest, agreed to consider the matter and six months later issued its report.

Though it rejected the claim that the cochlear implant should now be regarded as experimental – it was simply too late – its report greatly pleased the French Deaf community. The CCNE wrote that doubts regarding the precise benefits of the device were unlikely to be resolved in the near future. To avoid the possibility of compromising children's psychological and social development they should all be offered sign language from an early age, whether they might subsequently become candidates for implantation or not.

Despite the status of the CCNE, this report had little or no impact on actual practice in France. This of course is precisely an instance of the 'empirical slippery slope' that I referred to earlier: the notion that a consensus among ethicists may not be sufficient to change practice. However what was not possible in the USA or in France had already happened in Sweden. How, and why, and with what consequence?

In Sweden Gunilla Preisler (Professor of Psychology at the University of Stockholm) and her colleagues were also developing a highly critical reading of the evaluative literature. Not only were the studies almost all done by the implant teams themselves, and so were hardly objective, but most research was based on long-outdated theoretical assumptions regarding language development. The Swedish National Board of Health and Welfare agreed that the existing research literature was of little help given the circumstances prevailing in Sweden. Almost all the evidence for the effectiveness of the implant had come from countries in which oral education of deaf children dominated. They concluded, correctly, that in most countries the cochlear implant is intended to facilitate deaf children's access to spoken language (and to regular schooling). As we've seen, this is how its benefit is typically assessed. In Sweden its fundamental purpose is different. Swedish deaf children are educated using sign language, and the cochlear implant is to facilitate bilingualism: that is to say, the deaf child's access to spoken and written Swedish as a second language. If this is its purpose then almost all foreign research, starting as it does from another premise, and focusing on another outcome, is irrelevant to the Swedish situation. So Preisler and her colleagues began their own study of the psycho-social development of implanted children in Sweden. The results are quite the opposite of the conventional wisdom. Whilst the implanted children in bilingual settings were developing signing and oral skills, and also social skills, those in mainstream settings were too isolated from social interactions to be able to develop in a normal manner.

This example shows us a number of things. First, the overt endpoint in most studies, hearing and speech, derived from the expertise of the professionals (ENT surgeons, audiologists) who had jurisdiction over the technology. Economic analyses, often commissioned by a manufacturer, offered no challenge to clinical consensus. Second, even where –

as in France – an alternative (ethical) forum for assessment could be invoked, in which a wider range of evidence might be deemed relevant, this had no effect on practice. And third, the example of Sweden shows that only under quite distinctive political, social and cultural conditions could an alternative mode of evaluation, a distinctive endpoint, acquire legitimacy and authority.

Logically, we might say, endpoints and strategies <u>ought</u> to derive from social policies (as in the Swedish case above), or from the availability of healthcare resources, or from epidemiological circumstances. Our analysis hitherto suggests, however, that the universalistic and rationalistic aspirations of scientific medicine, combined with the difficulty of shifting to any sufficiently authoritative alternative jurisdiction, severely limit the possibilities of assessments sensitive to national difference. A second example highlights the interaction between professional and economic interests in sustaining this situation.

Endpoints, evidence and institutional commitments

Since 1960 two kinds of polio vaccine have been in general use: Salk's killed vaccine, which is injected and usually known as IPV and Sabin's attenuated virus vaccine, taken orally and known as OPV. For four decades the relative merits of the two vaccines have been debated. Although IPV came first, in the course of the 1960s it was the OPV that replaced it in virtually all countries. Production of IPV declined dramatically and rapidly: in the USA to zero.

Initial preference for the live vaccine derived from the central objective, which was of course to provide protection against infection. But this endpoint, protection against disease, proves to have a number of distinctive aspects. OPV was believed to confer longer-lasting protection. It was believed to be quicker acting: so offering more rapid protection (immunity being achieved in a matter of days rather than months). That meant it could be used in the event of a local epidemic. And thirdly was the argument that OPV provides protection to the community as a whole and could indeed be a route to eradication of the virus. The OPV, it was believed, provided an active infection of the bowel (whereas IPV does not) which would also interfere with the spread of polio through fecal matter and sewage. Attenuated live virus, excreted, would help protect those who had not been vaccinated. By 1964 the Committee on Control of Infectious Diseases of the American Academy of Pediatrics was writing that evaluation 'reveals a clearcut superiority of the OPV from the point of view of ease of administration, immunogenetic effect, protective capacity, and potential for the eradication of poliomyelitis'.[36]

There was, however, a cloud on the horizon. By summer 1962, with millions of doses of OPV having been administered in the USA, the suspicion emerged that in a small number of cases the attenuated virus in the vaccine had become virulent and itself caused polio. The Surgeon-General of the United States set up a committee of investigation. Careful analysis suggested that a few cases of polio were probably due to the vaccine itself.

In 1960 arguments in favor of the live vaccine had seemed irrefutable, but within just a few years matters had become more complex. Choice for one vaccine or the other now entailed weighing the presumed benefits of OPV (greater acceptability, community protection and so on) against what were now known to be small but definite risks associated with its use. Were the risks acceptable, and should society take them? Posed in this way, the issue is a political one in a very fundamental sense. If public health authorities are now obliged to re-weigh relative risks against relative benefits, we might expect a number of countries to change course. After all, there's no reason to suppose that so political a process would lead to the same result in each case, given wide variations in the remaining incidence of disease, coverage, resources, and so on.

By 1970, although incidence of the disease had fallen dramatically both the USA and Britain were still faced with hundreds of cases per year. In the developing world, of course, matters were far more serious. By this time the WHO had thrown its weight in the balance. The OPV was included in the Extended Programme on Immunization introduced throughout the developing world. Making matters complicated was the fact that three small countries, Finland, the Netherlands and Sweden, had continued to use the Salk vaccine alone, and with great success. How was the example of these countries to be balanced against what were still believed to be the valid arguments in favor of the attenuated vaccine? It is not difficult to imagine that the potential advantage of herd immunity – protecting the unvaccinated through indirect means – would be more persuasive the greater the distance still to go. In other words, where there were still hundreds of cases annually, and where vaccination levels were only 60 to 70 percent (as in the USA), the possibility of herd immunity could seem a more important advantage than where only five or ten cases occurred annually and vaccination levels exceeded 80 percent (as in Sweden or the Netherlands).

In early-1977 the Institute of Medicine published an analysis of the situation. A major issue is the small but politically significant risk associated with use of the attenuated vaccine. Between 1969 and 1976, 132 cases of paralytic polio had been reported in the USA, 44 of which were classified as resulting from the vaccine itself. Related to the numbers of doses of vaccine used, or the numbers of people vaccinated, the risk is estimated at one in anything between 4 and 23 million depending on the way risk is calculated. 'Such a risk would be acceptable,' Dr Nightingale (the project study director) writes, 'except that countries using only IPV report no serious complications'.[37] Did it thus make sense for the USA to abandon the attenuated (Sabin) vaccine in favor of the inactivated (Salk) vaccine? European countries using only IPV had managed to protect their populations without this risk of vaccine-attributable disease ... but countries like Sweden and the Netherlands had vaccinated more than 80 percent of their populations. This was not the case in the USA.

The evidence could be read as showing the superiority of the OPV, or of the IPV, or as suggesting the need for some intermediate strategy entailing the use of both vaccines. In fact, the tendency was to interpret it in ways that justified the *status quo*. In the event the virtually complete consensus around the OPV was not threatened. Few American or British experts were willing to take what seemed like the risk of recommending a switch back to the Salk vaccine. In the USA, the Institute of Medicine's new assessment of the relative risks and benefits of the two vaccines did not lead to the change in policy that the Institute recommended. Another 'empirical slippery slope' ... But now we can see better how and why this works.

The relationship between evidence-based argument and socioeconomic process had changed. Around 1960 it had been the arguments in favor of the OPV which had initiated an initial commitment by both public health authorities and by manufacturers. This kind of iterative convergence has been called 'lock in' by economists. A decade later the evidence no longer played the determinant role it had played previously. Now we see institutional commitments, the mechanisms underlying 'lock in', coming into play. Existing immunization schedules, established routines of healthcare workers, the familiarity and faith of the public: all of these rendered any change politically difficult. For the vaccine manufacturers investments tied up in existing facilities were a major issue.

It would, for example, be costly for a drug company to develop new manufacturing facilities and train people to make the Salk vaccine. Eli Lilly & Co, which used to make Salk vaccine, estimates that a

$30 to $50 million investment would be required over 3 years time. There is also some doubt that there would be an adequate supply of monkey kidney cells, which are used to grow the viruses for both vaccines but which are needed, some say, in greater quantity for the Salk vaccine than for the Sabin vaccine.[38]

This does not mean that evidence and argument have become unimportant. Far from it. Debate regarding the relative merits and uses of the vaccines is still conducted in the same terms as before. It remains necessary to justify any strategy in terms of the vocabularies of biomedical science and of public health. But the evidence, and the arguments, no longer determined policies and practices in the way that they had before. With the exceptions of Netherlands, Finland and Sweden, attenuated vaccine was in virtually universal use. Though concerns at the possible risks of attenuated vaccine had clearly arisen, these concerns were not strong enough to force any country seriously to reconsider its commitments to the OPV.

One exception, the Netherlands, was to prove important. The Dutch National Institute of Public Health (RIV), a public sector institute under the Ministry of Health, had actually developed a means of more efficiently producing a higher potency, standardized IPV, on a scale sufficient for the country's needs. There was little interest in exploring the possibilities of (re)developing an international market for IPV, at least to start with.[39] So how did the enhanced IPV reach international markets? Put in another way, through what processes was the technology exported from its local niche? Here evidence, the reflexive level, once more plays an important role. But there is a difference from the earlier period. Not only is collecting evidence for an alternative vaccine problematic by the late-1970s, given virtually global lock out, but at the same time it is very difficult to make the evidence convincing in the face of widespread scepticism. People did not want to be convinced that there might be advantages to the IPV.

The small group that set about re-establishing the value of IPV was a small and cohesive one, of which Jonas Salk, the French industrialist, philanthropist, and campaigner for world health Charles Mérieux, and Hans Cohen (Director General of the Dutch RIV) were leading members. Convincing the international public health community that IPV had a place in combating polio was not going to be easy. The argument that, unlike the OPV, there was no risk (of reversion to virulence) would not be enough. Data establishing the efficacy and duration of protection would also be needed. Cost was also an issue. To be more precise, it was necessary to show that there was some formulation of the enhanced IPV which was at least as good as the OPV under a wide range of conditions, but most importantly in tropical countries. To this end clinical trials would have to be conducted in Africa. Setting up those trials was to prove a difficult matter. They were eventually carried out in Francophone Africa, but only thanks to the political contacts of Mérieux and his colleagues in that part of the world.

Despite conclusions from these trials that the costs of IPV could be brought down, and that eIPV could even be more cost-effective in tropical countries, little changed in the late-1980s and early-1990s.[40] By the mid-1990s things were starting to change, at least in the USA, but that is another story.

Recall that on the basis of the cochlear implant example I concluded that Sweden's distinctive social policies implied, logically, a distinctive and very different assessment of the implant from that dominating the clinical literature. I also suggested that the universalizing and rationalizing aspirations of scientific medicine in practice seemed to stand in the way of evaluations reflecting national values, policies, and epidemiological circumstances. The history of the polio vaccines suggested that the process of 'lock in' also plays a role. In the Netherlands, the IPV had been built into a very effective national immunization programme, to which both public health and manufacturing practices had been adapted. Continuing to use the IPV made sense in the Netherlands, in a way that it perhaps wouldn't have in Britain or the USA. But attempts to establish the utility of the enhanced IPV in tropical countries, through clinical trials in Africa – according to those involved – posed a double challenge. On the one hand to the commercial interests of pharmaceutical companies looking to protect their markets; and on the other to the authority of the WHO. According to Philippe Stœckel, now of the *Fondation Mérieux*, the rejuvenated IPV threatened political and economic interests: 'we were bothering the WHO. We were an alternative, we were another solution. We were, they said, distracting people. With one goal, the use of OPV. We were sort of challenging them and they didn't like that'.[41] As Stœckel sees it, it is protection of their home markets by pharmaceutical companies with no IPV production facilities that is principally at stake here. Whatever academic advisory committees may say, whatever the epidemiological evidence, no practical role for IPV can be allowed.

The global politics of evidence

From the point of view of the medical profession establishing the utility of a new medical technology on the basis of standardized, quantitative assessments makes a good deal of sense. Objectification and quantification are a tried and tested response on the part of the professions to loss of public confidence in the subjective judgement of professionals.[42] The medical profession as a whole has a common interest in any analytic tool designed to provide 'objective' evidence for the effectiveness of what it does. If we focus down further we see how individual specialities work this out in relation to their specific and distinctive practices. Seeking control over a (new) technology that could have important implications for their work, authority and jurisdiction they try to control evaluation of the technology by imposing endpoints (their own definition of what should count as an indication that a technology works), and modes of data-gathering, derived from their own practices and conventions. That's what I found in earlier work on diagnostic imaging, and it is born out in the case of cochlear implantation.[43]

Sometimes, however, their specialist assessment of a specific technology may be subordinated to some higher jurisdiction. Such jurisdictions exist: the courts, legislatures, national ethical committees ... This of course is what happened in the United States, in the case of mammographic screening. Consideration of pediatric cochlear implantation by the French National Biomedical Ethics Committee was another example. In these non-specialist jurisdictions, more than just specialist interpretations of safety, efficacy and utility will be at stake. As other evidence, other discursive practices and considerations, begin to play a part, what seemed self-evident to the speciality may be that no longer. Before courts of law or legislative bodies not only will the strength of the evidence be differently interrogated, but very different evidence is likely to be considered in establishing the desirability of a new approach: for example, national laws and social policies, ethical standards, traditions, the social aspirations of minorities. This broad compass corresponds, of course, to the initial claims of Health Technology Assessment. The important thing is that speciality consensus might not be decisive enough. So now two issues arise.

First, how are issues brought before these alternative jurisdictions? Though forums available and access to them differ from country to country, it generally takes quite some social, economic or political capital to bring any medical technology before a non-medical jurisdiction. This might seem to be contradicted by the regulatory practices long in place for drugs, but now extended to other technologies. After all, drugs are assessed routinely and independently for safety and efficacy by bodies such as the FDA. As far as I can tell, the mandates and ways of working of such bodies are so construed as to exclude these broader considerations. For example, when I asked the FDA official responsible why the arguments of the deaf regarding sign language had played no part in their assessment of the pediatric cochlear implant, I was told that this went beyond their mandate.

Second, having got that far – what then? Even having succeeded in winning the argument, there is no guarantee that decisions will carry much weight. As the Deaf community in France found, the National Committee on Medical Ethics (one such forum) doesn't have the weight that the courts or the Congress have in the US. Its advice was not sufficient to change implantation practices in France. This is the 'empirical slippery slope' that I referred to earlier. The example of the polio vaccines throws light on what lies behind these 'empirical slippery slopes'. For there too, reports from, among others the National Institute of Medicine were, for a number of years, without effect on actual practice. As to what lies behind these processes, the answer seemed to be institutional commitments of the kind sometime called 'lock in' or 'path dependency'. Gradually, we saw in the case of the polio vaccines, it was less the evidence and the arguments that prevailed than the strength of existing commitments. Public health authorities were committed to established routines and did not wish to risk disturbing popular trust in those routines. Manufacturing industry had invested in production processes and did not wish to risk its investment.

Like the pharmaceutical industry, the medical profession also has an interest both in standardized and universalized assessments, and in sustaining the authority of its jurisdictions as well. Here there is a clear correspondence with Abraham's conclusions regarding processes of drug regulation, marked by harmonization and standardization of medical science aimed at increasing expert consensus. So long as scientific developments can be kept away from the awkward scrutiny of public interest groups and the mass media, then harmonization serves the interests both of medical scientists and the pharmaceutical industry, simultaneously.[44]

We have to recognize, with Abraham, that the assumption of universal validity, and global 'best choices' also corresponds to a vital interest of the pharmaceutical and medical devices industries, increasingly oriented to global markets and committed to the elimination of barriers to free trade. The political activity of these industries at the European level, the 'corporatism' that Abraham sees as marking European drug regulation, is an inevitable consequence.

But so what? What's the problem? In the case of drug regulation the adequacy of protection provided to consumers is an obvious and legitimate concern. The Thalidomide debacle of the 1960s is demonstration enough.[45] But what's at issue here isn't quite so tangible. Central to this paper have been two examples of countries going their own way. We saw how and why Sweden set about making its own assessment of the cochlear implant. We also saw how and why the Netherlands continued to use (and further improved) the IPV, despite virtually universal preference for the OPV. Should we regard these as positive examples? If so, if we believe that preferred solutions to health problems should be informed by respect for national differences in cultures, social policies, and patient preferences (at least in so far as these are not the result of socio-economic inequalities), then we are obliged to get to grips with the dynamic interplay of the national and the global. To deal adequately with this goes far beyond the scope of this paper. So let me restrict myself to one or two suggestions.

I have referred to the growing influence accorded patients/consumers. Their experiences and aspirations – what they bring to the discussion – will reflect social and cultural specificities. Will this provoke a shift in evaluative practices? Will their influence lead to changes in the forum within which technologies are judged? Logically, it should. But will it be allowed to happen? Will mandates and practices be sufficiently modified? The history of HTA is hardly reassuring in this regard. For the medical profession it could seem too much of a threat to an authority increasingly grounded in appeal to the 'scientific' basis of practice. For manufacturing industry too, a great deal is at stake.

Consider a well-known example from a slightly different area: the US lawsuit involving the European Union's ban on the sale – in E.U. territory – of beef treated with growth hormones. The WTO ruled in favor of the United States, categorizing this ban as an unfair protectionist practice that went against free trade, forcing the European Union to either allow the importation of these products or face severe sanctions, in spite of the opposition of a great majority of the continent's population. As one critic puts it 'The opinion of a few experts, chosen by the WTO authorities dealing with conflict resolution, thus overruled the democratically expressed wishes of the people of the European Union. In this case it was determined that the fear of consuming beef treated with growth hormones lacked scientific basis: inside the new world order defined by the WTO, this preference was not one for which people could legitimately opt'.[46]

The scope for national specificity, reflecting consumer values and preferences would seem to be threatened. Not only by the free trade doctrines associated with the WTO, but also by other more subtle notions of globalization. Recall how the WHO was unhappy at FAIR's attempts at re-validating the IPV, according to Stœckel and others, not because of commercial interests, but because the action seemed to undermine the notion of health problems having standard, global solutions. Preference for standard, universally applicable solutions seems to me to derive from interests shared between global health institutions, multinational industry, and the medical profession itself.

The argument of this paper leads, prospectively, in two directions.

One is epistemological. The medical profession is deeply committed to the scientific justification of its practices. Is it possible for this claim to be sustained whilst yet abandoning its present commitment to universality of assessments? I think it is. Popular opinion notwithstanding, there are many ways of being scientific. Scientific rigor can as well be demonstrated by agreeing on an approach, a methodology, rather than on an answer. This is what the sciences of the particular, history above all, surely have to teach. The implications of this must still be worked out.

The other is sociological. How can any such epistemological shift be brought about, given the economic as well as professional interests in the *status quo*? Where is pressure for change to be found, or created? Perhaps here the answer may be sought in the work of the cultural critics of globalization, like Appadurai, and the theorists of scientific citizenship.[47] Perhaps we have to find new approaches to the organization and conduct of health research, associated in some (yet to be established manner) with the growth of transnational health advocacy movements (such as, in the HIV/AIDS field GNP+).[48] A challenge, now, to what Appadurai terms the 'research imagination'.[49]

Notes

1 H. Marks, *The Progress of Experiment: Therapeutic reform in the United States 1900–1990* (Cambridge: Cambridge University Press, 1997); T. Pieters, 'Marketing Medicines through Randomised Controlled Trials: The case of interferon,' *British Medical Journal*, 317 (1998), 1231–3.

2 M. Berg, 'Turning a Practice into a Science: Reconceptualizing postwar medical practice,' *Social Studies of Science*, 25 (1995), 437–76.

3 R. Doll, 'Controlled Trials: The 1948 watershed,' *British Medical Journal*, 317 (1998), 1217–20.

4 J.B. McKinlay, 'From 'Promising Report' to 'Standard Procedure': Seven stages in the career of a medical innovation,' *Millbank Memorial Fund Quarterly*, 59 (1981), 374.

5 M.C. Creditor and J.B. Garrett, 'The Information Base for Diffusion of Technology: Computed tomography scanning,' *New England Journal of Medicine*, 297 (1977), 49.

6 H.D. Banta and S. Perry 'A History of ISTAHC: A personal perspective on its first ten years,' *International Journal of Technology Assessment in Healthcare*, 13 (1997), 430–53.

7 A. Faulkner '"Strange Bedfellows" in the Laboratory of the NHS? An analysis of the new science of health technology assessment in the United Kingdom,' in M.A. Elston (ed.), *The Sociology of Medical Science and Technology* (Oxford: Blackwell, 1977).

8 H. Bastian 'Speaking up for Ourselves: The evolution of consumer advocacy in health care,' *International Journal of Technology Assessment in Health Care*, 14 (1998), 430–53.

9 Alzheimer's Society, *Appraisal of the drugs for Alzheimer's disease* (London: The Alzheimer's Society, 2000).

10 S. Oliver, R. Milne, J. Bradburn, P. Buchanan, L. Kerridge, T. Walley and J. Gabbay 'Involving Consumers in a Needs-led Research Programme: A pilot project,' *Health Expectations*, 4 (2001), 18–28; B. Hanley, A. Truesdale, A. King, D. Elbourne and I. Chalmers, 'Involving Consumers in Designing, Conducting and Interpreting Randomized Clinical Trials: Questionnaire survey,' *British Medical Journal*, 322 (2001), 519–23.

11 S. Epstein, *Impure Science* (Berkeley: University of California Press, 1996); M.K. Anglin, 'Working from the Inside out: Implications of breast cancer activism for biomedical policies and practices,' *Social Science and Medicine*, 44 (1997), 1403–15.

12 D. Sarewitz, *Frontiers of Illusion: Science, technology and the politics of progress* (Philadelphia: Temple University Press, 1996).

13 D. Rothman, *Beginnings Count: The technological imperative in American health care* (Oxford and New York: Oxford University Press, 1997).

14 D. Callahan, *False Hopes: Overcoming the obstacles to a sustainable affordable medicine* (New Jersey: Rutgers University Press 1999), p. 57.

15 C. Pope, 'Resisting Evidence: The study of evidence-based medicine as a contemporary social movement,' *Health*, 7 (2003), 267–82.

16 J. Busfield ' Globalization and the Pharmaceutical Industry Revisited,' *International Journal of Health Services*, 33 (2003), 581–605.

17 J. Abraham 'The Science and Politics of Medicine Regulation' in M.A. Elston (ed.), *The Sociology of Medical Science and Technology* (Oxford: Blackwell, 1997).

18 L. Hogle 'Introduction: Jurisdictions of authority and expertise in science and medicine,' *Medical Anthropology*, 21 (2002), 231–9.

19 P.A. Kaufert, 'Screening the Body: The pap smear and the mammogram,' in M. Lock, A. Young and A. Cambrosio (eds), *Living and Working with the New Medical Technologies* (Cambridge: Cambridge University Press, 2000), pp. 177–80; J. Wells 'Mammography and the Politics of Randomised Controlled Trials,' *British Medical Journal*, 317 (1998), 1224–9.

20 S. Jasanoff, 'Science and the Statistical Victim: Modernizing knowledge in breast implant litigation,' *Social Studies of Science*, 32 (2002), 37–69.

21 W. Gardner 'Can Human Genetic Enhancement be Prohibited?' *Journal of Medicine and Philosophy*, 20 (1995), 65–84.
22 E. Parens (ed.), *Enhancing Human Traits* (Washington, D.C.: Georgetown University Press, 1998).
23 N. Holtung, 'Human Gene Therapy: Down the slippery slope?' *Bioethics*, 7 (1993), 402–19.
24 N. Daniels, 'The Human Genome Project and the Distribution of Scarce Medical Resources,' in T.H. Murray, M.A. Rothstein and R.F. Murray Jr. (eds), *The Human Genome Project and the Future of Health Care* (Bloomington: Indiana University Press, 1996), 173–95.
25 S.S. Blume 'Cochlear Implantation: Establishing clinical feasibility, 1957–1982,' in N. Rosenberg, A.C. Gelijns and H. Dawkins (eds), *Sources of Medical Technology: Medical innovation at the crossroads* (Washington, D.C.: National Academy Press, 1995), 97–124
26 E. Owens and D. Kessler (eds), *Cochlear Implants in Young Deaf Children* (Boston: Little, Brown & Co, 1989).
27 D. Kessler, 'Present Status of Cochlear Implants in Children' in Owens and Kessler (eds), *Cochlear Implants*, p. 218.
28 *Ibid.*, p. 220
29 A.Q. Summerfield and D.H. Marshall, *Cochlear Implantation in the UK, 1990–1994* (Nottingham: MRC Institute of Hearing Research, 1995).
30 National Institutes of Health, 'Consensus Development Conference Statement: Cochlear implants in adults and children,' May 1995, at http://consensus.nih.gov/1995/1995cochlear.
31 M. Haggard, Abstract 63, *Proceedings of the First European Symposium on Paediatric Cochlear Implantation* (Nottingham, 1992).
32 R. Carter and D. Hailey, 'Economic Evaluation of the Cochlear Implant,' *International Journal of Technology Assessment in Health Care*, 15 (1999), 520–30.
33 A.K. Cheng *et al*, 'Cost Utility Analysis of the Cochlear Implant in Children,' *Journal of the American Medical Association*, 284 (2000), 850–6.
34 H. Lane, *The Mask of Benevolence* (New York: Alfred Knopf, 1992).
35 S.S. Blume 'The Rhetoric and Counter-rhetoric of a "Bionic" Technology,' *Science Technology & Human Values*, 22 (1997), 31–56.
36 F.C. Robbins, 'Polio-historical,' in S.A. Plotkin and E.A. Mortimer (eds), *Vaccines* (Philadelphia: W.B. Saunders, 1988), 98–114, p. 104.
37 E.O. Nightingale, 'Recommendations for a National Policy on Poliomyelitis Vaccination,' *New England Journal of Medicine*, 297 (1977), 249–53.
38 P. Boffey, 'Polio: Salk challenges safety of Sabin's live-virus vaccine' *Science*, 196 (1977) 35–6.
39 Interview with P. van Hemert, Bilthoven, Netherlands, 25.3.1999 (Ingrid Geesink).
40 J.P. Moulin-Pelat, M. Garenne, M. Schlumberger and B. Diouf 'Is Inactivated Poliovaccine More Expensive,' *The Lancet*, 332 (1988), 1424.
41 Interview with P. Stoeckel, Marnes-la-Coquette, France, 29.1.1999 (I. Geesink and S.B.).
42 T. Porter, *Trust in Numbers: The pursuit of objectivity in science and public life* (Princeton, NJ: Princeton University Press, 1995).
43 S.S. Blume, *Insight and Industry: The dynamics of technological change in medicine* (Cambridge Mass: MIT Press, 1992).

44   J. Abraham and G. Lewis, *Regulating Medicines in Europe: Competition, expertise and public health* (London and New York: Routledge, 2001).
45   See for example H. Sjöström and R. Nilsson, *Thalidomide and the Power of the Drug Companies* (Harmondsworth: Penguin, 1972).
46   E. Lander 'Eurocentrism, Modern Knowledges, and the "Natural Order" of Global Capitalism,' *Nepantia*, 3 (2002), 245–68.
47   M. Elam and M Bertilsson, 'Consuming, Engaging and Confronting Science: The emerging dimensions of scientific citizenship,' *European Journal of Social Theory*, 6 (2003), 233–51.
48   GNP+, the Global Network of People Living with HIV/AIDS, can be found at www.gnpplus.net.
49   A. Appadurai, 'Grassroots Globalization and the Research Imagination,' *Public Culture*, 12 (1999), 1–19.

Index

Printed in the United States
By Bookmasters